Make: technology on your time

Volume 06

Features

76

BEAM ROBOTS
The BEAM approach to robot design creates nimble robots from simple components, with no programming required. Many BEAMbot behaviors are controlled by "nervous nets." These are simple control circuits that emulate low-level peripheral nerves in a spinal column.

40

PROTO
"Balloons provide a unique observation platform for doing planetary science that you simply can't do any other way," says Jeffery Hall, who leads JPL's aerobot research. "One of the great attractions of balloons is that it's fairly easy to make them. Then you get to take them outside and fly them around."

ON THE COVER
When MAKE Art Director Kirk von Rohr took these solar-powered robots (provided by solarbotics.com) outside for a photo shoot, they wouldn't sit still. Like a Hollywood cockroach wrangler, von Rohr was busy keeping the electronic insects from wandering. Cover photographer Douglas Adesko wisely confined the critters to an indoor setting.

Columns

Make: Projects

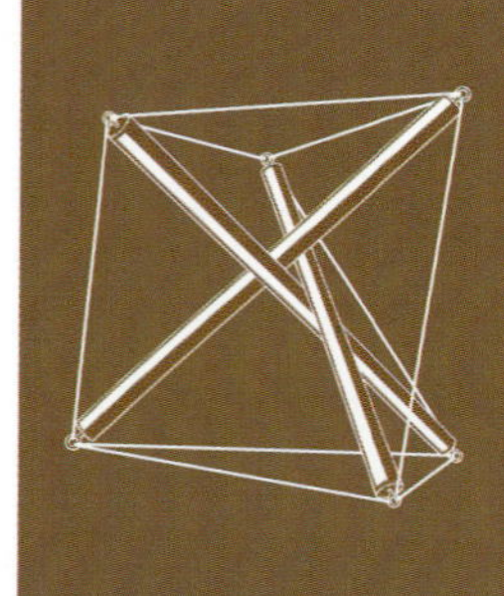

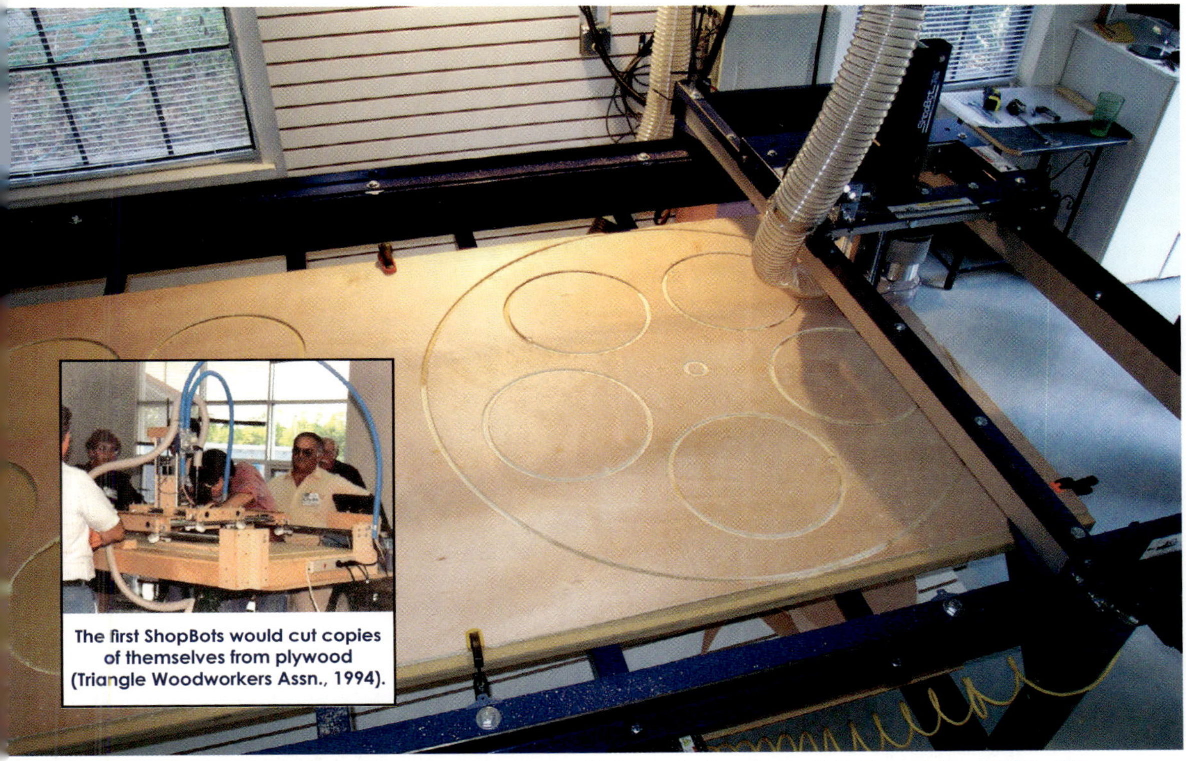

Volume 06

Maker

READ ME: Always check the URL associated with each project before you get started. There may be important updates or corrections.

117

MAKER PROFIILE
Steven Roberts' super-modified boat, christened *Wordplay*, looks more like a starfighter out of a science fiction TV show than something meant for the water. This analogy isn't too far off, considering the onboard technologies it packs: satellite and cellular phone, ham radio, and marine VHF.

EDITOR AND PUBLISHER
Dale Dougherty
dale@oreilly.com

EDITOR-IN-CHIEF
Mark Frauenfelder
markf@oreilly.com

MANAGING EDITOR
Shawn Connally
shawn@oreilly.com

SENIOR EDITOR
Phillip Torrone
pt@makezine.com

PROJECTS EDITOR
Paul Spinrad
pspinrad@makezine.com

EDITOR AT LARGE
David Pescovitz

STAFF EDITOR
Arwen O'Reilly

COPY CHIEF
Goli Mohammadi

COPY EDITOR/RESEARCH
Keith Hammond

CREATIVE DIRECTOR
David Albertson
david@albertsondesign.com

ART DIRECTOR
Kirk von Rohr

DESIGNER
Sarah Hart

PRODUCTION
Gerry Arrington

ASSOCIATE PUBLISHER
Dan Woods
dan@oreilly.com

CIRCULATION DIRECTOR
Heather Harmon

ADVERTISING COORDINATOR
Jessica Boyd

MARKETING & EVENTS COORDINATOR
Rob Bullington

ONLINE MANAGER
Terrie Miller

MAKE TECHNICAL ADVISORY BOARD:
Gareth Branwyn, Joe Grand, Saul Griffith, William Gurstelle, Bunnie Huang, Tom Igoe, Mister Jalopy, Steve Lodefink

PUBLISHED BY O'REILLY MEDIA, INC.
Tim O'Reilly, CEO
Laura Baldwin, COO

Visit us online at makezine.com
Comments may be sent to editor@makezine.com

For advertising and sponsorship inquiries, contact:
Dan Woods, 707-827-7068, dan@oreilly.com

MAKE is printed on recycled paper with 10% post-consumer waste and is acid-free. Subscriber copies of MAKE Volume 06 were shipped in recyclable plastic bags.

Please note: Technology, the laws, and limitations imposed by manufacturers and content owners, are constantly changing. Thus, some of the projects described may not work, may be inconsistent with current laws or user agreements, or may damage or adversely affect some equipment.

Your safety is your own responsibility, including proper use of equipment and safety gear, and determining whether you have adequate skill and experience. Power tools, electricity, and other resources used for these projects are dangerous, unless used properly and with adequate precautions, including safety gear. Some illustrative photos do not depict safety precautions or equipment, in order to show the project steps more clearly. These projects are not intended for use by children.

Use of the instructions and suggestions in MAKE is at your own risk. O'Reilly Media, Inc., disclaims all responsibility for any resulting damage, injury, or expense. It is your responsibility to make sure that your activities comply with applicable laws, including copyright.

Vol. 06, May 2006. MAKE (ISSN 1556-2336) is published quarterly by O'Reilly Media, Inc. in the months of February, May, August, and November. O'Reilly Media is located at 1005 Gravenstein Hwy. North, Sebastopol, CA 95472, (707) 827-7000. SUBSCRIPTIONS: Send all subscription requests to MAKE, P.O. Box 17046, North Hollywood, CA 91615-9588 or subscribe online at makezine.com/offer or via phone at 866-289-8847 (U.S. and Canada), all other countries call (818) 487-2037. Subscriptions are available for $34.95 for 1 year (4 quarterly issues) in the U.S. Canada: $39.95 USD; all other countries: $49.95 USD. Application to Mail at Periodicals Postage Rates is Pending at Sebastopol, CA and at additional mailing offices. POSTMASTER: Send address changes to MAKE, P.O. Box 17046, North Hollywood, CA 91615-9588.

Contributors

Lee Zlotoff (*MakeShift*), screenwriter for the original *MacGyver* series, has been making things since he realized both his hands were attached to the same brain. "No question, the inspiration for the MacGyver character came from cellar sojourns with my father, who could pretty much make *anything* work." He lives on a hill in Topanga, and when he isn't surfing or designing furniture and fountains, he makes television and film for Hollywood. Most recently he's launched his own media production outfit, Custom Image Concepts. "The muse said it was time to rock the boat. Who am I to argue?"

Danny O'Brien (*Life Hacks*) lives in San Francisco with his wife and daughter. By day, he fights crime at the Electronic Frontier Foundation as its activism coordinator. By night, he also fights crime, but only by neglecting to get around to it.

Matthew Russell (*Lego Soccer project*) is a Mac zealot, a skydiver, and a Krispy Kreme addict (he once ate 26 in one sitting). He graduated from the Air Force Academy in 2003, is a computer scientist by trade, and has 450 skydiving jumps under his belt. An inhabitant of the DC Metro area, he's a frequent contributor to O'Reilly's Mac DevCenter and likes to dabble with open source software for Mac/Linux. He has two pet ferrets, loves a freshly mixed black and tan from the tap, and always has too many irons in the fire.

Howard Wen (*Maker profile*) is obsessed with the weather — fretting over how hot it is or how cold it is. That's because he's from Dallas, where it's not unusual for the temperatures to change as fast as it takes you to read this paragraph. For his profile on tech-nomad pioneer Steven Roberts, Howard braced himself for the dreary gray skies and chilly rain of the Seattle winter — but the weather was sunny and warm when he visited. So he fretted about how hot it was.

Roy Doty (*Aha! illustration*) has been a cartoonist since the fourth grade. A degree in fine arts and a stint in the U.S. Army during WWII didn't distract him, and he still draws nonstop. Named Illustrator of the Year six times by the National Cartoonists Society, he has illustrated 176 books (29 of which he wrote), and his drawings have been published in so many magazines that the list would fill this entire page. His "Wordless Workshop" for *The Family Handyman* is in its 50th year. His favorite big lie? "That I'm 45 years old." His proudest achievement? "That I have never had a job for a single day in my life."

Contributing Writers:
Tim Anderson, Joost Bonsen, Gareth Branwyn, Mark R. Brown, Shawn Carlson, Bill Coderre, Travis J.I. Corcoran, Larry Cotton, Cory Doctorow, Nick Dragotta, George Dyson, Dan Gonsiorowski, Graffiti Research Lab, Joe Grand, Saul Griffith, William Gurstelle, Tom Igoe, Brian Jepson, Steve Johnson, Haje Jan Kamps, Peter Kirn, Mike Kuniavsky, William Lidwell, Matt Lind, Greg Lipscomb, Merlin Mann, Mister Jalopy, Annalee Newitz, Quinn Norton, Danny O'Brien, Tim O'Reilly, Ross Orr, Tom Owad, Dave Prochnow, Michael H. Pryor, Joel Raedeke, Michael Rosenblatt, Tyler Rourke, Matthew Russell, Bob Scott, Liam Staskawicz, Bruce Sterling, Damien Stolarz, Dan Strunk, Robert Bruce Thompson, Howard Wen, David Williams, Tom Zimmerman, Lee D. Zlotoff

Contributing Artists:
Douglas Adesko, Jake Appelbaum, Nathan Arnold, Kevin Bain, Scott Beale, Ricardo Blanc, Roy Doty, Nick Dragotta, eboy, Mike Fields, John Granen, Alex Handy, Dustin Amory Hostetler, Harold R. llano, Timmy Kucynda, Chris Leschinsky, Tim Lillis, Christopher Lucas, Dave McMahon, Jim Mullins, Visa Parviainen, Joe Reinhardt, Seth Schoen, Damien Scogin, Shawn Sinyork, James Sooy, Robin Tafel, Travis Thatcher, Jay Townsend, Susan Williams

Interns:
Adrienne Foreman (web)
Jake McKenzie (engr.)
Ty Nowotny (engr.)

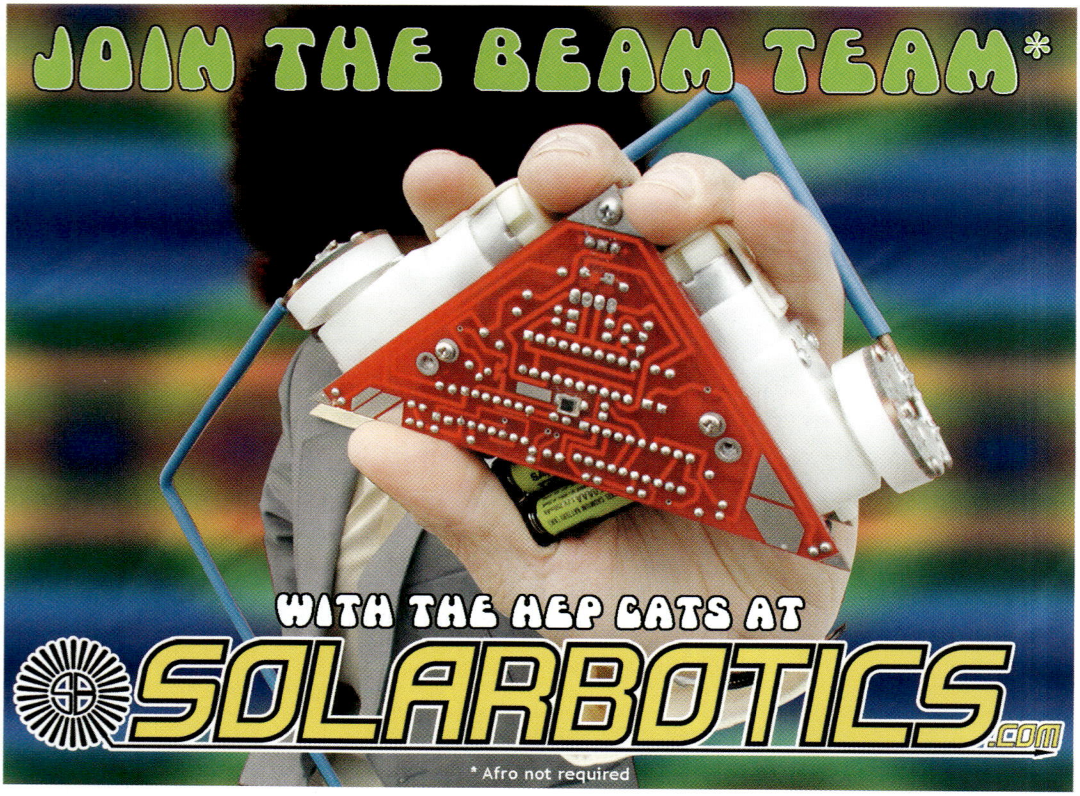
JOIN THE BEAM TEAM*
WITH THE HEP CATS AT
SOLARBOTICS.com
* Afro not required

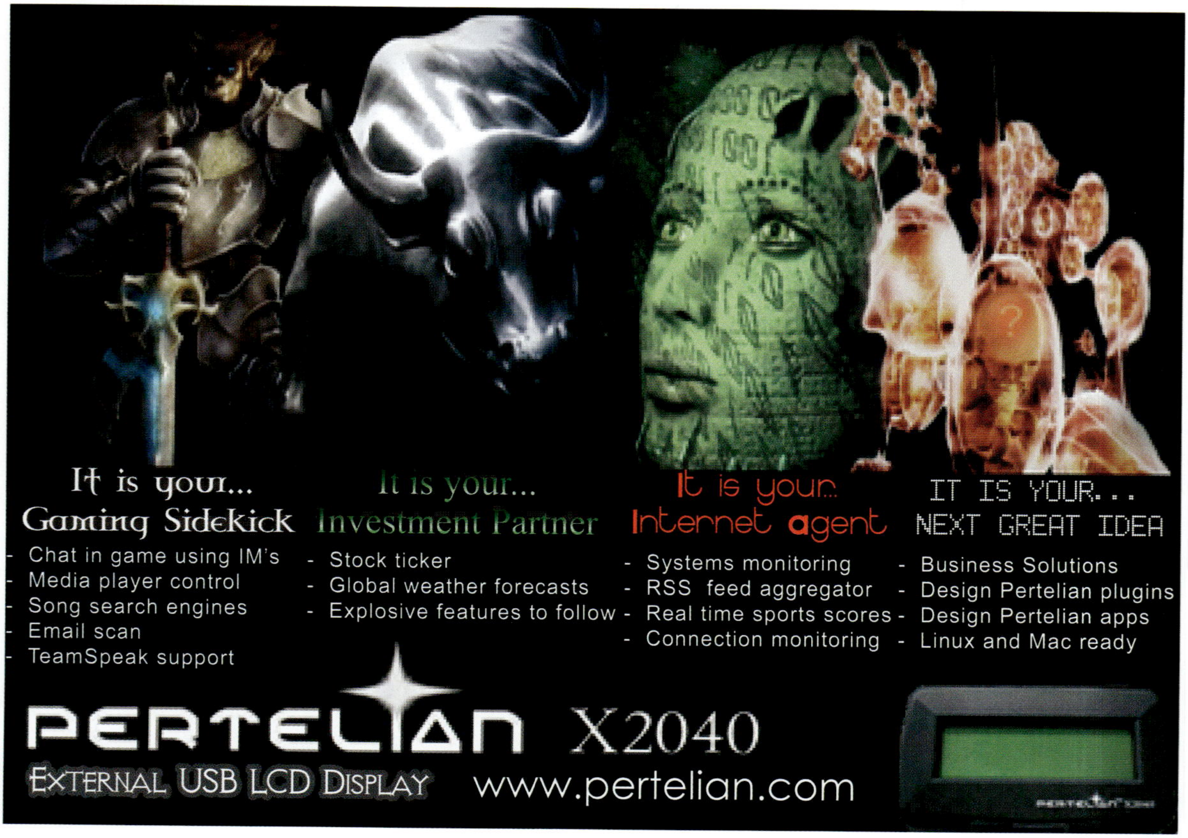
It is your...
Gaming Sidekick
- Chat in game using IM's
- Media player control
- Song search engines
- Email scan
- TeamSpeak support
It is your...
Investment Partner
- Stock ticker
- Global weather forecasts
- Explosive features to follow
It is your...
Internet agent
- Systems monitoring
- RSS feed aggregator
- Real time sports scores
- Connection monitoring
IT IS YOUR...
NEXT GREAT IDEA
- Business Solutions
- Design Pertelian plugins
- Design Pertelian apps
- Linux and Mac ready
PERTELIAN X2040
EXTERNAL USB LCD DISPLAY
www.pertelian.com

Dale Dougherty

TONES DEM TONES, DAMN RINGTONES

AT THE WEB 2.0 CONFERENCE, A PANEL of teens talked about their habits and attitudes for a wide range of gadgets and services. All of them said they almost never pay for digital music, and seemed incredulous that anyone would. Then the moderator asked them about cellphones. One of the teens confessed that he spent $50 a month on ringtones. FIFTY DOLLARS A MONTH! A red light on my maker radar went off.

Shouldn't people make their own ringtones, not buy them? They already own the music. Don't they know how to get it on their cellphones?

The business press is happy to repeat the phrase, "the blah-blah billion-dollar ringtone market worldwide" (the number they say is $4 billion, but when they use big, unsubstantiated numbers so freely, what I hear is "blah-blah"). This means lots of people are making money on ringtones, and they do so mostly by pickpocketing teens and their parents.

Imagine if websites could automatically charge your ISP for activities that your children engage in online. Ringtone vendors, which include carriers, phone manufacturers, and third-party vendors, are taking advantage of the automatic billing relationship behind every cellphone. Increasingly, the game is to get teens to initiate recurring monthly fees hidden from parents in a large, complicated phone bill.

One San Diego parent has brought a class action suit against Jamster, charging fraud and false advertising. This Verisign subsidiary offers a "free" ringtone and then enrolls anyone who asks for it in a service costing $5.99 a month. They advertise on MTV, Nickelodeon, and the Cartoon Network, also promising a free ringtone if you send them a text message. What they don't advertise is that they automatically sign you up for their costly service.

Another infuriated father set up a webpage (www.sleaze-mobile.com) to rant about what happened once he noticed that his 11-year-old daughter had been duped into a $3.99-a-week ringtone service. His screenshots show how a teen could easily think she is saying yes to a free download, since the message saying that she is actually signing up for the service is on a later screen.

Why would anyone pay more for a 30-second ringtone than for a song on iTunes? One answer is that teens see ringtones as personalizing their phones while music downloads are just pure entertainment. Isn't this a terribly shoddy view of personalization?

Do you personalize your phone by signing up for a $5.99 monthly plan to download "I'm N Luv (Wit A Stripper)" by T-Pain, the number one ringtone on Jamster today? Or by downloading Beyoncé's "Check On It" from Cingular's Media Mall for $2.49, where you can't get to the song without seeing promos to "Idolize My Phone" by downloading "Yo Dawg" and other "famous and fun sayings" from *American Idol*?

And why do teens buy so many ringtones? Apparently, teens assign a distinctive ringtone to each friend. The more friends, the more ringtones you buy. Or is it the more ringtones you buy, the more friends you have? It's a sign of your social network.

Of course, DIY ringtones are the best option for true personalization, but don't expect to find good information in your phone's manual or on your carrier's website. They don't want you to know how.

You *can* own your own ringtones, and we'll do our best on makezine.com to cover the tools and techniques for making them on your own.

Dale Dougherty is editor and publisher of MAKE.

Tim O'Reilly

NEWS FROM THE FUTURE

WE HAVE MET THE ARTIFICIAL INTELLIGENCE, AND HE IS US.

REMEMBER THE 1970S TV SHOW, *The Six Million Dollar Man*? It featured Lee Majors as an injured astronaut rebuilt with technology that made him faster, stronger, and more capable. He was a cyborg, a fusion of man and machine.

Anyone with prosthetics — even someone with eyeglasses — is a cyborg of sorts, but what really excites the imagination is today's news of direct connection between brain and machine.

Last fall, University of Florida researcher Thomas Demarse grew a culture of 25,000 rat neurons, then watched as it taught itself to control an F-22 flight simulator. In February, Klaus-Peter Zauner at the University of Southampton, U.K., hooked up a slime mold to a six-legged robot. The biologically controlled robot scrabbles away from bright lights, emulating the behavior of the mold. Last year, a monkey with a brain implant learned to control a robotic arm with its thoughts. In human clinical trials, paraplegic Brian Nagle has succeeded in playing video games with a similar implant.

If it's possible for neurons to control a machine, it should also be possible for machines to control the brain. DARPA researchers recently presented a plan to remotely control hammerhead sharks via a neural implant. The hope: stealth spies able to track enemy ships.

But what has put bionics and the man-machine interface on my radar lately isn't just these modern echoes of old science fiction, but rather the idea that the latest web applications are forging a more subtle, but no less profound, merger of man and machine.

Boxxet founder You Mon Tsang recently introduced a new meme into my vocabulary: *bionic software*. Boxxet is an example of such a system. It's a collaborative news site (a kind of "my.digg.com") in which customers can put Boxxet's spiders to work collecting data on any subject — and then site users rate the results to train the spider. The result is a tool that harnesses both computers and humans to deliver better results than either can do alone.

When You Mon described Boxxet in this way, it struck me that this "bionic" aspect is critical to many of the most successful web applications. Back in 2003, I began using an illustration of von Kempelen's *Mechanical Turk* in my talks, to emphasize the point that one of the things that distinguishes web applications from PC-era applications is the fact that web applications actually have people inside them, working daily as part of the application. Without the programmers running the crawl at Google (and updating the anti-spam algorithms), without the users feeding the spiders by continuously linking to new sites, the application stops working. In a profound way, the users are part of the search engine. This turns out to be true in one form or another for almost every breakthrough web application.

I generalized this idea into one of the key principles of Web 2.0, namely that Web 2.0 applications are systems for harnessing the collective intelligence of their users. But the term *bionic systems* gives a new twist on this concept.

I was talking about the idea of bionic software with venture capitalist Tom Shields a few weeks ago, and explaining how I thought old dreams of artificial intelligence are being replaced by this new model, in which we are creating more intelligent systems by using humans as components of the application. Shields neatly summed up the paradigm shift: "AI becomes IA" (Artificial Intelligence becomes Intelligence Augmentation). We have met the AI, and he is us.

Now that I understand that we're building a next generation of bionic systems, I'm seeing them everywhere. I'd love your thoughts. Where else are you seeing this fusion of human and computer to build capabilities beyond the reach of either alone?

Check makezine.com/06/nff for related stories.

Tim O'Reilly (tim.oreilly.com) is founder and CEO of O'Reilly Media, Inc. See what's on the O'Reilly Radar at radar.oreilly.com.

DEATH BEFORE DULLNESS!

HOW TO DO BORING TASKS, OR WHY ROBOT-RUN DYSTOPIAS CAN BE FUN.

By Merlin Mann and Danny O'Brien

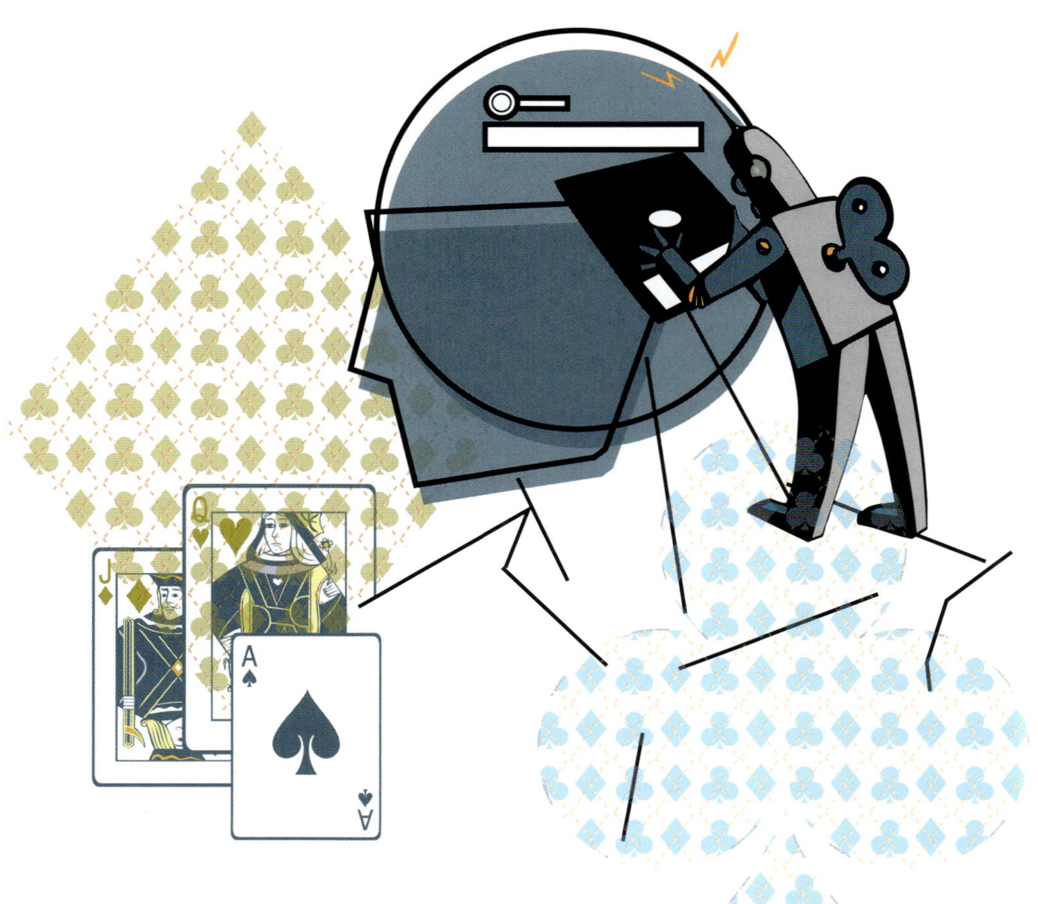

Illustration by Sarah Hart

I THINK MANY LIGHTLY ORGANIZED GEEKS suffer from an overpowering fear of tedium. It's part of being a novelty-seeking kind of person: when you get substantial amounts of pleasure from uncovering new knowledge, any task where you're not learning — where you're just *repeating* — is unbearably painful to pursue.

If you're a programmer, there's an additional level of suffering in drudgery. Doing the same thing over and over while coding, like copying and pasting the same function into different functions? That's never a demonstration that you're doing good work. It's usually a sign that something is wrong with your code. You're paid to remove redundancies. Programmers look at a task like weeding the lawn and think, "I've already done this. Why can't I wrap it in a loop, and just run that?"

As a result, geeks often get themselves into terrible trouble because they cannot bear to do simple, everyday tasks — like pay bills on time, or fill in a tax form, or wash the dishes. It's an aesthetic and moral offense. Better, says your subconscious, to be bankrupt and on the run from the IRS, eating from plastic cups and plates, than to have to do some brain-numbing piece of repetitious nonsense, even for a moment. Death before dullness!

Imagine my surprise, then, as I researched super-organized geeks and I discovered how many apparently mind-numbing routines they indulge in. They have scripts, sure, and little automated hoo-hahs, but they also perform certain repetitious actions, no matter how easy it might be to optimize them away.

The repetition serves to keep certain facts fresh in their minds. Plenty of alpha geeks have a complex system of repeated behaviors for filing a new document or creating a blog entry. Even when they keep their to-do lists on the computer, they still monkishly copy it out onto paper first thing in the morning. The ritual of filing or blogging helps them remember what they're saving. Copying the to-do list fixes the tasks in their minds, and helps to keep the list small.

My favorite story regarding this skill is from the legendary coder John Carmack, creator of *Doom* and *Quake*. Carmack went to Las Vegas for a break and, Vegas being Vegas, decided to have a wager. Carmack being Carmack, he beat the house. The game he played was blackjack, and the technique he used was card counting — which lets you beat the house odds as long as you persistently follow a strictly defined set of rules. As he wrote:

"Playing blackjack properly is a test of personal discipline. It takes a small amount of skill to know the right plays and count the cards, but the hard part is making yourself consistently behave like a robot, rather than succumbing to your gut instincts."

Consistently behaving like a robot is the challenge of many everyday tasks. The hard part for many smart people is to stop thinking, and start acting as though you're running a program *on yourself*.

For many of us, that goes against our guttiest of instincts: we feel as though we're deliberately making ourselves more stupid. But as a strategy, sometimes you really don't want, or need, to be smart.

Many geeks use games to soften the blow. They take boring, monotonous tasks, and by turning them into a challenging game, eke some tiny squib of enjoyment from them.

The games you play depend on your personality. If you're naturally competitive, you'll want to have some external measure to constantly attempt to beat. If you're someone who enjoys music, match your actions to an internal, complicated rhythm.

Stop thinking and start acting like you're running a program *on yourself*.

I'm about as competitive as anyone always picked last in gym, and music is all bleep bleep bleep to me. But I've gotten through many a terrible job by tapping into the rich bureaucracies of science fiction and thriller shows. Programs like *CSI*, *Battlestar Galactica*, and *Star Trek: The Next Generation* gain some of their appeal from the occasional glimpse of their heroes conducting a tremendously boring task, or minions milling around busily in the background. This leaves opportunities to place you in that world, only temporarily stuck doing the dull Space Taxes Forms.

This can work in almost any otherwise depressing scenario. As science fiction author Rudy Rucker once said: "For me, the best thing about cyberpunk is that it taught me how to enjoy shopping malls, which used to terrify me. Now I just imagine the whole thing is two miles below the moon's surface, and half the people's right brains have been eaten by roboticized steel rats. And suddenly it's interesting again."

Learn how to reel in your mind at Danny O'Brien's lifehacks.com and Merlin Mann's 43folders.com.

Bruce Sterling

HANDS ON: ELEGANT INNOVATION

FROM LAMPS AND CHAINSAWS TO SAVING THE PLANET.

CERTAIN AREAS OF THE WORLD HAVE an innately MAKE-like approach to life. Northern Italy (where I'm writing this installment of my column) is one of those places. Case in point: famed Milanese architect Michele De Lucchi.

For the Italian designer outfit Artemide, De Lucchi created the legendary Tolomeo work lamp. The Tolomeo is bright, sleek, and ductile, with long columnar metal arms and tendon-like wire. It silently bends and swivels at a touch, and stays poised in any position you may place it in. Since its creation in 1983, the Tolomeo has been the number-one work lamp that designers themselves buy for design labor. It's the designer's designer lamp and has been selling merrily for decades.

Oddly enough, De Lucchi is not a designer. He's a "radical architect" from the 1970s, when young Italians rebelled at the constrictions of their discipline and exploded laterally into postmodern home décor, weird laminated bookcases, couture, electronics, graphics — in a word, most anything hackable. This eclectic approach has many practical benefits.

At Olivetti, where De Lucchi worked for 20 years, he involved himself in the production of some 240 products, finishing his career there as the corporation's creative director. He also designed door pulls, tape dispensers, laptops, chairs, vases, interior décor for banks and hotels, and much more.

So far, so good: we're describing a world-famous, multitalented Italian designer at the top of the profession. Now comes the really interesting part: explaining why De Lucchi spent much of 2005 making conceptual art with chainsaws.

Explaining the fondness for chainsaws comes easily enough. To make his point, De Lucchi produces a cherished 12-year-old Italian fountain pen from his immaculate jacket. It occurred to him that although pens and pencils are used with great grace, precision, elegance, and tenderness, no one has extended this approach toward the humble, industrial chainsaw. Why not? Are chainsaws less worthy than pencils? There is no alternative to the presence of industrial objects in modern life.

A large tree fell near De Lucchi's home. That incident required a chainsaw. This was a chance to learn. Once he had his goggles and gloves on, De Lucchi knew that the chainsaw had been radically underexploited as a means of creative expression. The 55-year-old maestro soon made it his business to own and master a variety of chainsaws.

Like many architects, De Lucchi spends much of his professional life making small-scale models of housing. So he decided to refine his chainsaw skills by making model homes straight from the dead tree. No fussy stickler for mere handicrafts,

> "Every project is a voyage from idea to realization. There is an ocean of compromise in the middle."

De Lucchi also added telling model details with a laser cutter and a water jet.

These chainsawed model homes look like they were whittled into shape with a giant's jackknife, but the unique models sold at once to eager art collectors. Some of the models were botched. Those, he discarded and wrote a book about: *Twelve Stories About Little Houses*. These chainsaw failures were too ugly to show in public or to display as art, but the effort to make them taught him useful lessons. This resulted in a good set of design war stories.

"Every project is a voyage from idea to realization," he tells me in careful English. "There is an ocean of compromise in the middle."

The architectural lessons from the chainsawed homes are now reflected in De Lucchi's ambitious Japanese eco-village development, outside Osaka. This is a big effort, an entire Japanese suburb, but he has learned, he says, to seek his inspiration for

Photograph by Bruce Sterling

These chainsawed model homes look like they were whittled with a giant's jackknife.

bigness in that which is small, simple, and intuitive. Big, corporate research-and-design teams are all very well in their place, but they are big by nature, and concerned with big resources. So, they are always anxious to avoid big mistakes.

One cannot experiment properly in a state of anxiety. Creativity is closed off by fear. It's even worse to fail to be anxious at a big scale. It's wrong to arrogantly experiment with the lives and fortunes of a company's employees and stockholders — as if those many people didn't matter.

By their nature, big companies and mass production will "commercialize, marketize, banalize, and globalize." But if industry is to improve the world, industry needs something truly good to work on.

Therefore, De Lucchi has divided his own work into sets of physical scales. First, there are the small things he does alone in a home office: "experiments, searches, and fun." By design, these efforts have no deadlines, no clients, no deliverables, no budget, and they are done without commitment to anybody.

At the next level comes a small company called Produzione Privata (Private Production). This atelier features De Lucchi himself, his design assistant, a bookkeeper, and a producer, whose job it is to outsource the manufacturing of De Lucchi's designs. Produzione Privata is deliberately small, but it sells real products and it has a real budget. The next and final step is the De Lucchi architecture firm, aMDL, which does large-scale urban work in Germany, Russia, Japan, Italy, and elsewhere.

These different levels of creative scope do not conflict. Instead, they support and refresh one another. One level is no more or less "serious" than the next. They are a creative ecosystem, where the scale and muscle of the bigger firm can shelter the little greenhouse of the new, and where the small innovative experiments can provide a unique edge and unheard-of innovations for the bigger outfit.

"There is no alternative to industrial organization," says De Lucchi. "But we must also believe that we have the chance to reach a better world through industry. An industry is more than a public investment. If man believes in industry, but industry fails to believe in humanity, the planet is finished."

I don't know about the planet, but having met De Lucchi, I know that Milan and he are the polar opposite of "finished." They have found the means, motives, and opportunities for elegant innovation.

Bruce Sterling (bruce@well.com) is a science fiction writer and part-time design professor.

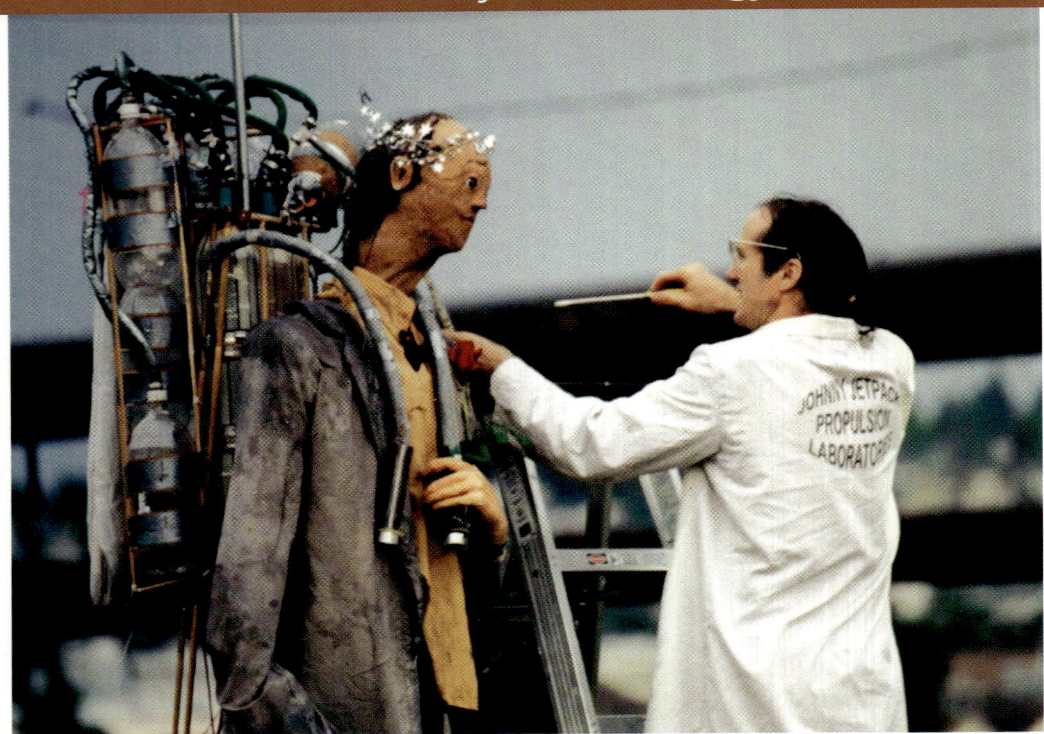

Johnny Jetpack

It's difficult to match the sophisticated and subtle humor of throwing a dummy off something really high and watching it fall to the ground, but to be fair, it's a low-tech laugh.

Luckily, **Nathan Arnold** of Seattle has discovered that strapping a dummy to a compressed-air jetpack, launching it more than 100 feet in the air, and then watching it fall to the ground is three times as funny and provides some geek appeal on the side.

Arnold's performance-art dummy show and his eternal desire to maximize mayhem led to the idea of launching the thing — Johnny, by name — as high in the air as possible in 1983. "It turned out to be more spectacular than I thought it would be," he says, and after more than 100 launches, he hasn't stopped laughing.

Johnny looks surprisingly substantial on the launch pad, so much so that you'd never guess that he weighs in at a svelte 8 pounds. The jetpack — a hodgepodge of carbonated beverage bottles and tubing strapped to a bamboo frame — tips the scales at 30 pounds, but the majority of that weight is liquid accelerant, which will be expelled forcefully within a second or so of launch.

"The hardest part is getting a good seal," says Arnold. "Once you have a seal, you're 90% there."

After Johnny and the jetpack are set up, pointed skyward, and a good seal allows the bottles to be charged with water and compressed air, Arnold has only to pull a pin to initiate the launch.

Johnny accelerates rapidly once the jetpack is activated. His head, filled with cattails, lolls forward from the force, and a Rube Goldbergian sequence of events is set in motion. As the liquid drains from the main chamber, the drop in pressure triggers the parachute ejection device. The removal of the parachute initiates the separation of Johnny from his jetpack, and an "explosion" of cattails blows Johnny's head off. More compressed air chambers in Johnny's chest cavity cause his arms to flail while a parachute deploys to return the jetpack safely to the Earth. Meanwhile, Arnold and whoever is lucky enough to be present laugh at it all.

Thirteen years ago, Johnny Jetpack made his first flight, and over 100 launches later Arnold refuses to stop tinkering on what will forever be a work in progress. You have to protect the technology, of course, so the decision over who gets the parachute — Johnny or the jetpack — is easy. But the jetpack is fragile, and even with a gentle landing, each launch requires roughly 40 hours of preparation. When that kind of time is involved, it's easy to write it off as a back-to-the-drawing-board situation every launch.

What's coming next? Water, compressed air, and soda bottles have served Johnny Jetpack well for a while now, but there comes a time when the laws of physics draw a line. Striding over that line, Arnold picks up a canister of liquid nitrogen. "It's a possibility I've been looking at," he says.

The MythBusters television guys tried to find a way to launch something more mortal than Johnny with 2-liter bottles, but without a good background, they failed despite mounting 15 bottles to a pack. Arnold thinks they just weren't trying hard enough.

"Improper line of force," he says, "They concluded that there was no way to launch a person with soda bottles. I don't agree with that."

—Dan Gonsiorowski

Nathan Arnold's website: johnnyjetpack.com

Photography courtesy of Nathan Arnold

Big Heads

"You maniacs! You blew it up! Ah, damn you! Damn you all to hell!" Oh, my mistake. This isn't a post-apocalyptic world where apes rule, but rather the parking lot behind artist **David Adickes**' studio near downtown Houston.

Adickes is from the bigger-is-better school of making. In addition to a 67-foot-tall statue of Sam Houston in Huntsville, Texas, and 43 busts standing 20 feet tall at Presidents Park in Williamsburg, Va., and the Black Hills of South Dakota, Adickes has big plans for a 36-foot-tall statue of the Beatles, a giant four-person bust of historic figures alongside a busy highway he'll call *Mount Rush Hour*, and a 280-foot-tall cowboy, which would be the largest statue in North America (the Statue of Liberty is about half as tall).

A young 79, Adickes fell in love with gigantism after visiting Mount Rushmore. But he soon found out that creating Rushmore-sized sculptures was no easy feat. He started Sam Houston in 1992 without really knowing how to pull it off.

"It was really a work of engineering without any blueprint," he says. So Adickes learned to apply a simple rule of thumb to making monuments of extreme proportions: "when in doubt, make it stout." And stout they are. His statue of Sam Houston uses five layers of concrete poured over steel mesh attached to a welded steel framework — over 30 tons of material.

Given the extra challenge of scale, why the fixation on gigantic statues? "What I am trying to prove," says Adickes, "is the same thing Frederic-Auguste Bartholdi, the creator of the Statue of Liberty, was trying to prove: think big, do it right, people will come. Today, 108 years later, the Statue of Liberty is an icon, and two million people take a boat to visit it every year. Two years after Bartholdi, Gustave Eiffel built a tower in Paris. Vehemently criticized, it later became the symbol of France."

Whether Adickes' works will endure the test of time like Bartholdi and Eiffel remains to be seen, but there is no question that his titanic figurines make for big attractions and big impressions — just follow the line of tourists. *—William Lidwell*

>> **List of related sites:** makezine.com/06/made

Photograph by William Lidwell

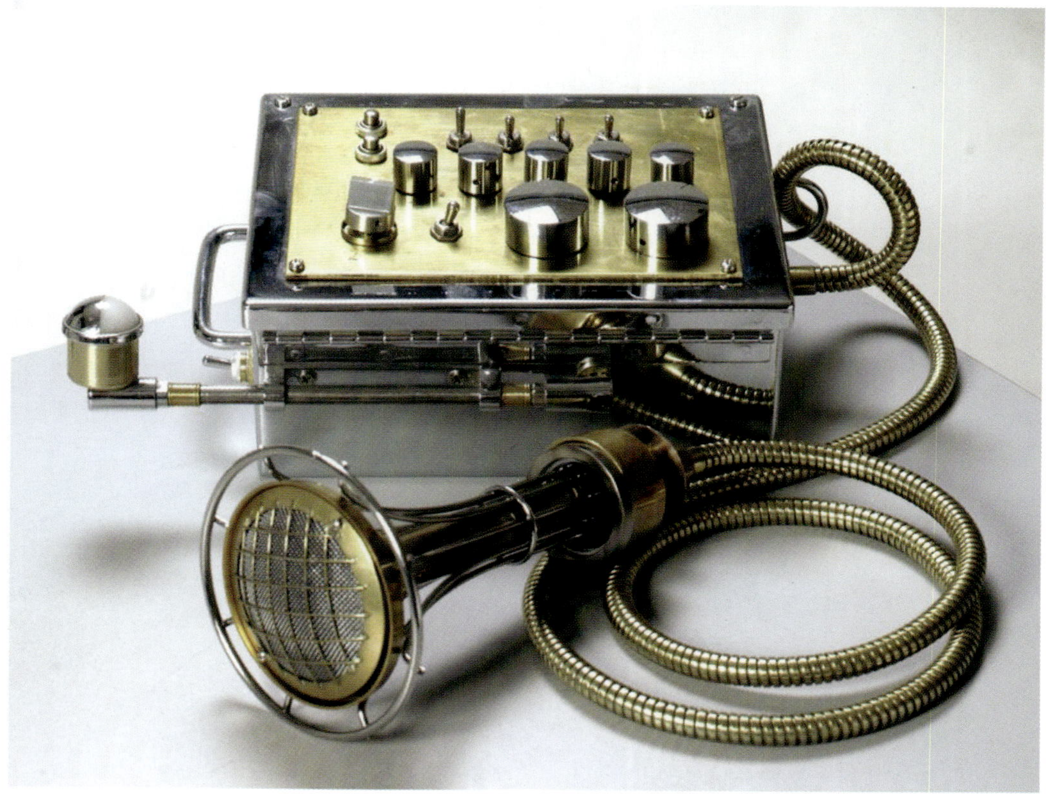

Sound Made Art

Beeps-a-Lot Box, *Arcane Device*, and *Stylophonic Device* each answer a question I hadn't considered: What would have happened if H.R. Giger had decided to teach band?

These graceful sculptures are working musical instruments produced by Gulf Coast artisan **Mike Ford**. Designed to evoke curiosity as well as admiration, each is heavy with mysterious controls, indicators, and attachments, all beckoning to be explored.

Growing up in a family of Gulf fishermen, Ford learned the value of inventiveness early on. Even as a tyke, he embraced this heritage, secretly designing and building a control panel for an imaginary rocket ship. Sadly, his space exploration dreams were cut short after some critical knobs had to be unglued and returned to his grandmother's TV set.

A skilled stringed instrument builder, Ford was inspired by an article on circuit bending to start modifying discarded electronics boards. Dissatisfied with common project boxes for these creations, he began developing cases as exotic as his circuits.

Finding this new challenge artistically satisfying, he pursued refinements to his design process and metalworking skills while an art student at the University of Mississippi.

Now a full-time artist, Ford melds the electronics of his instruments with cases that, despite their fluidly seamless look, are largely composed of repurposed or found pieces that he works with hand tools. His design is decidedly deco-industrial, a look that has captivated him since he first glimpsed the retro-futuristic gadgets in movies like *Dune* and *Brazil*.

This inspiration is easy to see in the wonderfully strange effectors on his instruments, whose functions seem both obvious and inscrutable, giving his gear the look of something from a not-quite-parallel world (a world where they use a lot of chrome).

Besides building up his stock of sculptures, Ford is now constructing microsynthesizers based on vintage analog integrated circuits. I'm guessing those won't be in plastic boxes. *—Bob Scott*

>> **Mike Ford's website:** mikefordsculptures.com

>> **His work is also featured at:** getlofi.com

Photograph by Kevin Bain

Fly Boy

Look! Up in the sky! Is it a bird? A plane? No, it's **Visa Parviainen**. Last October, Parviainen, sporting a birdman suit and custom jet boots, dove face first out of a hot air balloon high above Lahti, Finland, and took off into the wild blue yonder.

Parviainen is one of a growing number of sky divers who wear wingsuits during their dives. The fabric spanning the legs and arms enables the free-fallers to glide a bit before popping their parachutes. Boosted by his rocket boots, though, Parviainen was able to zip along at 2,000 meters for several minutes.

To build his jet-powered flight suit, Parviainen and his cohorts at BirdMan, Inc. attached a pair of off-the-shelf microturbines to a pair of old hockey skates. Fueled by kerosene, each engine spits out about 16kg of thrust. Tests at a nearby university's wind tunnel convinced them that the aerodynamics should work out. The trick was figuring out a fuel tank system that was lightweight and durable enough for a twisting, turning, windswept human body.

"The solution was to use hot water bottles as fuel tanks," Parviainen says. "Since they're flexible, it's also easier to squeeze every last drop of fuel out of them when you're flying."

Next, the team built a special launch platform to suspend from the side of the balloon canopy. The platform had two purposes: one, it kept exhaust from the boots away from the balloon and passengers as Parviainen revved the engines before takeoff. Two, it was a "nice lounge" for the ride up.

Parviainen knows his way around a machine shop, souping up cars and motorcycles for racing and hacking mounts for helmet cams and other skydiving gear. He's spent the winter working on a new rev of the flight apparatus, substituting a different set of microturbines and tweaking the engines for more reliable operation at chilly temperatures. On his next flight, he also plans to wear a black box recorder of sorts.

"Someday, I want to take off from the ground and land too," Parviainen says. "It's far in the future, but I do think it's possible. Right now though, this is all just good fun." *—David Pescovitz*

BirdMan: bird-man.com

Photograph courtesy of Visa Parviainen

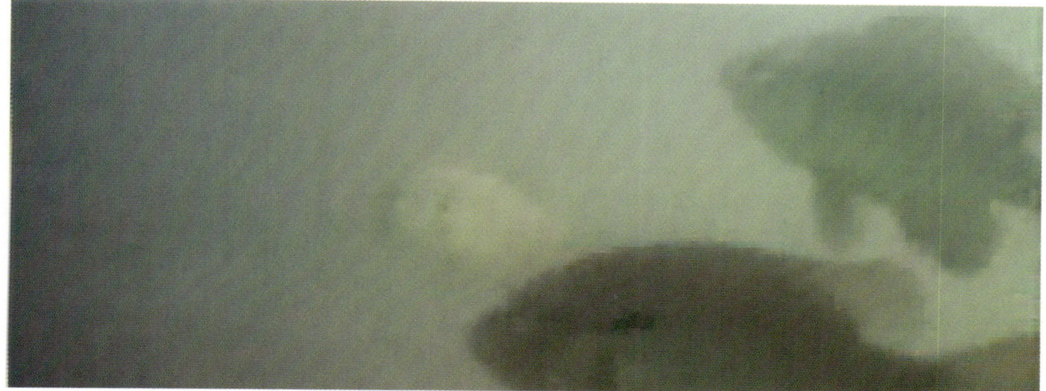

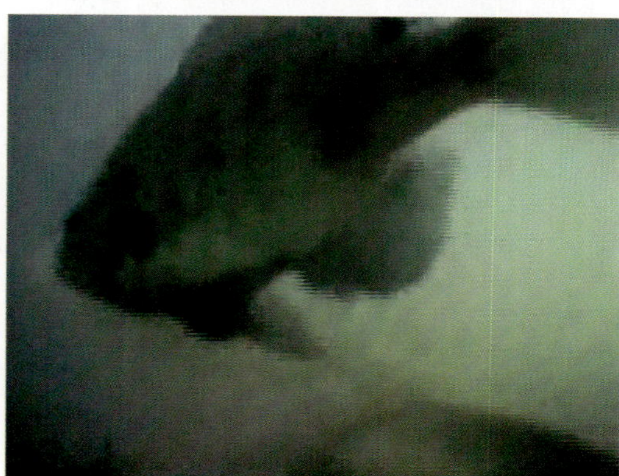

The Eye Aquatic

Cousteau and the undersea world. Cameron and *Titanic*. And now, **Joe Reinhardt** and **Mike Fields** and the depths of Lake Moraine in upstate New York.

Every summer, Reinhardt and brother-in-law Fields tackle a new DIY project at the family's lakeside camp. In 2005, the pair built their own underwater ROV (remote observation vehicle) with two video cameras feeding live images to a shipboard laptop — all for about 100 bucks.

Making things is second nature for Reinhardt, 24, a computer tech in digital imaging. This time, he got to indulge his underwater fascination: "I always wanted to be a marine biologist," says Reinhardt. "I love the water, and ships ... and watching Discovery Channel with the real ROVs exploring the *Titanic*."

Reinhardt and Fields built their homebrew ROV's frame out of PVC pipe, and its transparent camera housing out of scrap quarter-inch-thick acrylic tube from the local plastics supply (milled to watertight tolerances on a friend's lathe). They joined the two with simple but strong carpenter's ratchet clamps.

The B&W video camera was $29 from Harbor Freight, complete with infrared LEDs for night vision, power supply, and 80 feet of RJ11 cable. They scored a Chinese color "Spy-Cam" for $1 on eBay (plus $35 shipping), and ran its video signal up the audio wire in the RJ11 cable.

After an embarrassing misfire with ballast tanks ("We put 'em on top, so it sank upside down every time"), the explorers improvised a solution ("a big hunk of concrete and a bungee cord") and lowered their ROV to the lake bottom to capture video of sunfish, perch, and muskie sporting in the wild. The rig proved watertight to 40 feet.

This summer, they're going deeper: their 2006 model has thrusters for true independent ROV mobility, using watertight 12VDC motors coupled to propellers by super-powerful neodymium magnets. It'll be rated to 200 feet, good enough to dive quarries or wrecks on Lake Erie, Reinhardt says. James Cameron might want to check his rearview mirror.

—Keith Hammond

>> **Homebrew ROV photos and video:** makezine.com/go/ROV

Photography by Joe Reinhardt andMike Fields

Good Ship Popsicle Stick

He used to break bones. Now former Hollywood stuntman **Robert Mc-Donald** uses popsicle sticks to break world records.

McDonald has built three Viking-ship replicas out of ice cream sticks. All have been seaworthy, including his latest beast, built from 15 million popsicle sticks over three years. He's now working to break another record by sailing the ship across the Atlantic Ocean in true Viking fashion.

"I have a dream to show children they can do anything," he says. "If they can dream it, they can do it."

In fact, that's what started McDonald down this popsicle path — he wanted to encourage his 8-year-old son to aim high and believe he could succeed, all the while making the world a better place. He is adamant about creative recycling — all the ice cream sticks he used were previously used or imperfect, and were donated by the Ola ice cream company in Europe. McDonald's home port is in the Netherlands.

"[We're] demonstrating how amazing objects can be created from everyday, recycled goods," he enthuses.

"Creative" and "fun" pepper his conversations. And he lives what he speaks. In April 1986, McDonald rocked his way into the world record book by rocking in a chair for 340 hours. Last year, he grabbed another record by sailing a ship made from 370,000 ice cream sticks, the *Baby Ola Bison*.

The bigger replica is 50 feet long and weighs in at a hefty 13 tons, including more than two tons of glue. Named *Mjollnir* (mil-ner) — the Viking god of thunder — she is an open craft with no protection for the sailors whatsoever. The 6-person crew sleeps in true Viking style: hammocks strung across the deck. Her voyage across the Atlantic began in mid-April.

McDonald heads the Sea Heart Ship Foundation, a group spreading fun to kids in hospitals around the world. Captain Rob (as the kids call him) recently returned from a hospital tour of Florida, the Gulf Coast, and New Orleans, where he gave away 28,000 stuffed animals in 14 days (yet another record).

—Shawn Connally

>> Great Atlantic Crossing Adventure: obvikingship.com/index.php

Photography courtesy of the Sea Heart Ship Foundation

Photograph by Travis Thatcher

Duet with a Robotic Drummer

Gil Weinberg is having trouble with his drummer: he's trying to get both of Haile's arms to work at the same time. That would be an unusual problem, except for the fact that Haile is a robot.

Haile's microprocessor-controlled, motorized arms are able to play ordinary acoustic drums, with expressive control over timbre and dynamics. But while the movements of its anthropomorphic, wooden body are impressive, Haile's listening ability is as important as its playing. Using custom computer software developed in the Max/MSP multimedia environment, Haile can analyze the performance of a human drummer and respond in real-time.

"We've tried to create a new musical experience — to surprise you," says Weinberg. Haile's responses range from simple imitation to variation and even intelligent accompaniment. The results vary: sometimes, the algorithms simply don't work, or don't work in a way that makes sense to human ears. But Weinberg, who was a jazz pianist for years before he became interested in computers, is most excited by the moments at which Haile feels like an equal musical partner, and plays in ways no human would.

"I've played with people for years," says Weinberg. "But especially when you're in a particular genre, you know what to expect. Here, we are in uncharted territory."

While Haile's human-robot communication skills continue to evolve, it's also getting a chance to facilitate human-human communications. Next, Haile travels to Jerusalem to play on a program featuring collaborations between Jewish and Arabic percussionists. The composition is called *Jam'aa*, or "gathering" in Arabic. Drawing on the communal tradition of Middle Eastern music, Haile will interact with professional darbukah and djumbe players, transforming what they play. At least, once the remaining technical bugs get solved.

"Talk to me Monday," says Weinberg; by then, he hopes Haile will be able to play with both arms.

—Peter Kirn

Robotic Drummer: makezine.com/go/robotdrum

Cory Doctorow

A HOUSE DIVIDED

HOLLYWOOD FEELS THE STING OF ITS OWN COPYRIGHT LAWS.

FOR CASUAL USERS, ITUNES DRM doesn't look so bad, and for many of them, it's a pretty loose set of chains. A lot of info-civilians don't own more than five CPUs (the limit on the number of boxes an iTune can play on), and iTunes is new enough that not many of us have had the opportunity to sink a big investment in the technology and try to take it with us to a new vendor.

For the recording industry, iTunes represents a chance to charge more for less. An iTune can't be sold on as a used good, something that we can do with our CDs. You can't turn an iTune into a ringtone, nor make other uses that the record companies might want to charge us money for. It's a neat and easy way to gouge us for the stuff we used to get gratis.

But for Apple, iTunes is a great way to lock us into using its products. No one is allowed to make a compatible iTunes player without Apple's permission (when RealNetworks tried to make a Real player that ran on the iPod, Apple threatened a lawsuit under the DMCA).

Apple doesn't really care how many CPUs you listen to a track on. Apple doesn't really care how many people you stream a song to. That's just a sop to the record industry. Apple cares about you sinking an investment into an integrated chain of technologies (iPods, CPUs, iTunes metadata) and media (songs) that will raise the switching costs if you decide to give your business to someone else.

From the recording industry's perspective, this is a disaster. Once Apple controls the relationship with music listeners, Apple can name its price and set its terms for selling music. The record companies are trying desperately to raise the cost of some tracks and lower the cost of others because they believe they'll sell more that way, and Apple has basically laughed them off. Why? The record companies will never stop selling songs for use with iTunes. Apple controls the iPod, and uses that control to lock Rhapsody and Napster songs out of the most popular portable player on the market. The music industry can't afford to blow them off.

The entertainment industry, having accidentally created cozy monopolies for the likes of Apple, Microsoft, and Macrovision, is now busily trying to put itself back in charge of its own destiny. Next-generation crippleware DVD standards, like Blu-Ray and DVD-HD, are being defined by studio-crafted, fragile alliances of bitter competitors from the technology industry, companies that will be hard-pressed to present a unified front when it comes time to negotiate terms with Hollywood.

Once Apple controls the relationship with music listeners, Apple can name its price and set its terms for selling music.

While monopolists rearrange the deck chairs on their sinking superliners, we makers are busily building, acquiring, and using open technologies and file formats that let us move our media freely from one device to another, keeping us from being locked into an entertainment company or a DRM vendor.

Let the dinosaurs go for each other's throats — we can see the tar rising around their feet.

Cory Doctorow (craphound.com) is a science fiction novelist, blogger and technology activist. He is co-editor of the popular weblog Boing Boing (boingboing.net), and a contributor to *Wired*, *Popular Science*, and *The New York Times*.

1+2+3 T-Shirt Laptop Pouch

By Ross Orr

Reincarnated tee "iShirt" protects your computer and deters thieves.

You will need: Sewing machine, marker, pins, old T-shirt. A smaller T-shirt works best for this, unless your laptop is enormous. An old favorite with holes and burn marks will just add to the case's theft-deterring stealth. A pocket large enough to store your laptop's power brick is a nice touch.

1. Mark and pin shirt.

a. Lay the shirt face up and then fold the bottom half upwards over the chest. Check by touch that the bottom of the pocket (if the shirt has one) is a bit above the fold. Set your laptop on top of the shirt and then draw vertical marker lines about an inch from both of its sides.

b. Along each marker line, use straight pins every few inches to hold all the layers of fabric together. Pin crossways so that the pins won't get snagged on the sewing machine's foot.

2. Sew and cut excess.

a. Sew a straight seam along both marker lines. For reinforcement, double up the stitching near the open edge of the pouch.

b. Cut off the sleeves and trim the excess cloth around the seam, leaving about 2 inches all around. This extra fabric outside the seam will provide additional padding.

3. Use!

Turn the pouch inside out so that the pocket is on the front again. Slip your laptop into the middle compartment. The neck opening is great for stashing stray cables; the bottom flap holds papers or your ever-present copy of MAKE.

For Extra Credit

+ Take one of the leftover sleeves and sew it into a pouch for your iPod and earbuds!

Ross Orr hacks low-tech gadgets and invasive plants in Ann Arbor, Mich.

Illustrations by Dave McMahon

Maker

Tech-Nomading From Shore to Ship

Twenty-three years ago, he went on a bike trip and never returned.

INTERVIEW BY HOWARD WEN
PHOTOGRAPHY BY JOHN GRANEN

Steven Roberts' workshop is a mess. Nestled in the quiet woods of Camano Island, a small community situated by Puget Sound about 90 minutes north of Seattle, the 3,000-square-foot building is overflowing with miscellaneous electronics, computers, and gadgets. Tools are everywhere — on shelves, on worktables, on the floor.

System Documentation
AMPRO
Clippard
GENOA QUICK COUPLINGS
Microship Project
Fiberglass
Vorsicht: Laser in Betrieb
AMPRO
AMPRO
AMPRO
McMASTER-CARR
111

"There's an old saying: The smaller the boat, the bigger the adventure."

Roberts tells me he's in the process of eBaying most of these things. I notice a stack of Macintosh computers. They're the really old "classic" models, the ones with the tiny built-in black-and-white monitors. The genial Roberts, a tall and bearded fellow, says I can take one or more of them if I'm interested. I politely decline, being one of those people who avoids collecting items for which I have no use.

I'm a little surprised by all of this stuff in his place. When I first contacted him over the summer of 2005, Roberts told me how in 1983, then 31 years old and living in the suburbs of Columbus, Ohio, in a three-bedroom ranch-style house, he worked as a freelance writer (reporting on technology and electronics), and felt trapped. "I was working my ass off to pay for things I didn't want, a lifestyle I didn't want. I was doing things I didn't enjoy," he recalled.

Seeing his workshop, however, I was reminded of an email in which he lamented: "Almost everyone I know is bogged down by 'complexity': Creaky but familiar tools one doesn't dare replace; new toys not yet learned; incompatible power supplies; unlabeled mystery cables clogging drawers; lost documentation; and the vulnerability of it all becoming utterly useless."

Back in 1983, back with all those things he didn't want, Roberts yearned for simplicity and adventure. He had not planned on it, but he was about to become a pioneer of the "tech-nomadic" life: a man who used mobile technology to live on the road, to stay in touch with the world wherever he was, and to free himself from a dreary existence.

The Souped-Up Canoe

Roberts became interested in electronics when he was 8 years old and growing up in Louisville, Ky. His father was a mechanical engineer, and Roberts was inspired by him to build machines.

Roberts' road warrior box, the Shacktopus contains three transceivers, GPS, wi-fi, environmental telemetry, solar power, speech synthesis, audio recording, Bluetooth, a deployable antenna, and more.

As a child, he made numerous electronic projects, which he usually entered into the school science fair: an induction magnet that could pick up aluminum, a Morse-code translator, and a speech synthesizer based on his own vocal tract, taken from an X-ray of his head.

He shows me his latest project in his workshop, a souped-up canoe. Actually, "souped-up" is an understatement. It's been totally overhauled with the addition of two smaller hulls, one connected at each side to the main hull where the pilot sits. Eight blue solar panels, four on each side, are set across from the smaller hulls to the main hull. This water vehicle even has retractable wheels. Why does a boat need wheels? Roberts felt it would be more convenient to move from shore to water if it had wheels, so he spent years designing and building this elaborate mechanism.

This super-modified boat, christened the Wordplay, looks more like a starfighter out of a science fiction TV show than something meant for the water. This analogy isn't too far off, considering the onboard technologies it packs: satellite and cellular phone, ham radio, and marine VHF. And this doesn't even include the video cameras and other gizmos that aren't mounted to it the day I visit. There's a circular, triton-looking antenna set toward the craft's bow that Roberts explains is an "ultrasonic transducer." Once a second, each of the three forks communicates with the other in order to collectively measure the surrounding air mass, wind speed, and wind direction.

Yet Wordplay travels by decidedly low-tech means: by wind with a sail.

The Winnebiko

Roberts shows me around the rest of his cluttered workshop. One item catches my eye; it looks like a control panel ripped out of a jet fighter cockpit. It served as the control panel for his first tech-nomadic vehicle, the Winnebiko.

Back in Columbus, 1983, Roberts went to a party out in the country. That night, he stared into a campfire and then things "just all snapped

into place": why not combine the things he was most passionate about — computers, writing, travel, bicycles, and romance — into a new life?

He ordered a custom-built recumbent (a type of sit-down bicycle), then grafted on a RadioShack TRS-80 Model 100 laptop, a Hewlett-Packard HP-110 portable computer, CB radio, and a 5-watt solar panel to power these gadgets. He named the resulting vehicle the Winnebiko.

Starting from Columbus, Roberts biked 10,000 miles, passing through small towns along the southern East Coast, through Florida, the South, Texas, and the Southwest, and ending in Silicon Valley in California about 18 months later. Throughout this trip, he continued to earn a living by writing articles on his laptop. He also wrote about his journey, which eventually caught the notice of people in the media, who wondered about this man bicycling across America on a "computerized bike."

As fulfilling as this really long bike ride had been, he found it frustrating that he could not write while riding at the same time. "I had all this mobility, but I was just watching the words flow away, knowing by night, when I was camping or whatever, I would not capture these thoughts," he says.

Roberts upgraded the Winnebiko for an encore trip in 1986. The Winnebiko II added packet radio for email access, a security system with motion detection and voice synthesis, and a new, more sophisticated control panel. To enable himself to write as he pedaled, he hacked apart the keyboard of the TRS-80 Model 100, and rewired the keys to the bike's handle controls.

His second bike tour ran from Seattle, along the West Coast, and across the country to the East Coast. He was accompanied by his girlfriend at the time, Maggie Victor, who rode her own recumbent. Together, they traveled 6,000 miles.

Behold the BEHEMOTH

The attention Roberts got from the media led to interest from corporate sponsors. From 1988 to 1992, he threw more things onto the Winnebiko II. A lot of things. Much like a succeeding version of a Microsoft application, the bike quickly became bogged with too many features, to the point of absurdity and uselessness. So many components were put on it that a trailer had to be designed to hold them, and for the bike to tow.

Renamed BEHEMOTH (Big Electronic Human-Energized Machine ... Only Too Heavy), it was assembled in Silicon Valley by a team of volunteers assisting Roberts. It had almost every piece of mobile and computer technology at the time: a hacked Macintosh and other computer systems, tons of radio communications devices, GPS navigation, even a radiation monitor.

"I got so distracted by the tech stuff," Roberts admits. "I'd be reading a trade journal and go, 'Ooh, ooh, I could use that!' and then I would schmooze and get it."

The media and public were enchanted with his journeys, and wowed by the technology-laden, though impractical, bike. Roberts was interviewed by many reporters and appeared on TV talk shows. But as public interest in his project was reaching its peak, the tech-nomadic biker's passion for his original dream was dying. Ironically, to make public appearances and do speaking gigs, he traveled the country in a diesel truck that carried the BEHEMOTH in its trailer.

The Laboratory on an Island

Using the money he earned from his speaking tour for the BEHEMOTH, Roberts bought property on Camano Island, choosing the region for its variety of surrounding waterways. He put up the 3,000-squarefoot workshop to facilitate the research, construction, and testing of small watercraft that would utilize tech-nomadic technologies. He brought over the volunteer-community ethic of BEHEMOTH by inviting engineers and other specialists to take part.

The "Microship Project" began with the idea of outfitting a basic kayak with communications devices, but evolved into a pair of specially modified boats, Songline and Wordplay. These water vehicles expanded upon the embedded systems technologies that Roberts and his BEHEMOTH team developed, and which allowed the pilot to control almost every aspect of the craft through a Palm-OS PDA.

In his workshop, Roberts shows me the inside of a thick plastic project box that's sitting at one end of a worktable. It looks large enough to hold several hundred sheets of 8.5x11 paper. This is the power supply he designed for the Wordplay. It uses seven microprocessors to regulate and distribute 600 watts. While the boat's solar panels alone can provide 5 knots to move the craft, this power unit was also designed to enable the pilot to divert all power to the thrust for emergencies (like quickly veering away from another boat).

Roberts took the Wordplay on a 132-mile ride through Puget Sound in September 2001.

"Unlike the other systems [Winnebiko, et al.], which were really lifestyle choices, this is much more 'grab and go,'" he says. "With Shacktopus, I'm avoiding any more multi-year projects that tie me to a specific bike or boat."

He shows me the prototype for the Shacktopus, sitting on the same worktable where he has the power supply box for the Wordplay. The Shacktopus is essentially a clear plastic project box with a medley of off-the-shelf communication and other computer components inside,

"It was like I reached around the back of my head and hit the reset button," he says of the day his journey began.

Though he still considers it to be in development, the Microship Project has gone through extended inactive periods over the years, as personal priorities for him shifted.

"When I started this ten years ago, I was perfectly pleased with the idea of taking out on a canoe-sized hull, spending two years sleeping in the bilge — it's the size of a coffin. Now I'm 52, and I'm like, 'I don't want to be that uncomfortable for that long!'" he says, chuckling.

Project Shacktopus

Along with a weakening tolerance for physical discomfort, Roberts has been wondering lately if maybe he has spent too much time over the past 20 years designing machines and not enough of it going out on actual adventures. He calls this consequence the "BEHEMOTH effect."

Thus, last year he decided to create an all-in-one mobile communications pack, which he named the Shacktopus. He plans to make it the size of a notebook computer, or small enough to fit into a messenger bag. It's the first project of his that he hopes to turn into a commercial product.

bashed and interfaced together through a unified control system designed by Roberts. There's an HF/VHF/UHF transceiver; a GPS system; environmental telemetry; internet access with wi-fi bridging; a lithium-ion battery power system that can be charged by a solar panel, automobile cigarette lighter, or AC outlet; speech synthesis; audio recording; Bluetooth interfacing to a notebook computer or PDA; and a deployable antenna array from HF to 2.4GHz.

For something that's supposed to help simplify Roberts' life, to free him from dealing with such technically complicated projects as the Microships, this early version of the Shacktopus itself already looks to be ... complex.

While it stems from the restlessness that has been growing within its creator over the past few years, the Shacktopus also seems to represent the same conundrum. Roberts came to the woods of Camano Island with what sounded like a simple enough plan: build a boat and move on to the next big adventure. This didn't happen fast enough. Apparently, it takes a long time to build a boat, at least the way Roberts likes to build one, thanks to fancy things like retractable wheels. Things became complicated.

"I want to get moving again, and I see Shacktopus as the way of making that happen — kind of short-cutting that whole process," he says, perhaps hopefully. "I'm looking for a big boat that I can live on and do some world traveling."

The Allure of Human Power

When I started talking to Roberts for this story last summer, I immediately noted that his tech-nomadic vehicles — the bikes and the boats — shared the distinction of being small vehicles that relied mainly on people power. I wondered what the appeal in that was for him.

"I like the human scale of it. I find that when you cruise on a motorcycle or car, you're really anonymous. You're just somebody passing through on the freeway. Whereas, when I was on a bicycle, I was completely non-threatening," he said to me back then. "Back in the early 80s when I was [biking through] small towns in the South, people would take me home. They weren't worried about me. Also, it's a lot more satisfying. Kayaking to an island is more exciting. There's an old saying: the smaller the boat, the bigger the adventure."

Afterwards, I mused for several months: what could be the thematic connection of a human-powered vehicle, or wind- and solar-powered one (as in the case of the Microships), to mobile communications technology? What was the appeal of the two brought together?

When I meet the tech-nomadic pioneer in person on a sunny afternoon in February 2006, I ask him about this. Beyond the fact that the two subjects have always interested him personally, he cannot come up with some profound, satisfying explanation for it all.

I climb aboard the Wordplay — or climb into it, to be more precise. At 5'8", I'm much shorter and thinner than Roberts, so the inside doesn't feel "coffin-like" to me at all. There's a lot of legroom. I fiddle with some of the levers and try to relax myself into the hard seat. I look ahead, out the canopy. A compass is affixed to the top of the dashboard, and beyond that I see a large marker board with technical-looking diagrams drawn on it, hanging from the wall in front of the Wordplay. If I were on Puget Sound, my view instead would be of the water rippling out beyond and into a backdrop of the mountain ranges of northwestern Washington, I imagine. But it also feels like I'm in a starfighter.

It then gradually dawns on me: there's a unique feeling about being inside such a small craft that relies upon your own physical skills and wits to control. It becomes like an extension of your own skin, your own body. It becomes personal.

Maybe that's the connection: mobile communications, and vehicles like Roberts' bikes and this boat, both evoke a personal, emotional bond between themselves and the user. The more physically invested the rider is in the direct powering of the vehicle, the more personal the journey becomes.

"My goal is not to spend my life in the lab building electronics," Roberts says. "I got this beautiful place in the woods. It looks like it ought to be paradise, but I'm just itchin' to get moving again."

The last time he felt that itch, he was living in a three-bedroom, ranch-style house somewhere in the suburbs of Columbus. There was no eBay, so he was stuck with a bunch of unwanted stuff that he couldn't easily get rid of.

But one day, he left it all behind — there were still dirty dishes in the kitchen sink — and didn't even bother to lock the front door of his house. He just pedaled away on an odd-looking bike that he had slapped computers and mobile communications gadgets onto. "It was like I reached around the back of my head and hit the reset button," he says of that day when his journey began.

The microship Wordplay:
1. Retractable wheels facilitate transport.
2. The cockpit is covered with a bright red canopy made of heavy waterproof fabric.
3. Solar panels can provide thrust in an emergency, otherwise it's wind and sail.

1

2

3

Sky Cutters

Pulling the strings at the Berkeley Kite Festival.

By Arwen O'Reilly

IF YOU'RE IN BERKELEY IN JULY, YOU CAN'T miss the kites. After almost 20 years, the Berkeley Kite Festival is a time-honored tradition and one of the largest kite festivals on the West Coast, attracting kite-flying professionals, families with children, and casual passersby. Held on parkland surrounding the Berkeley Marina, you can see the kites — swooping, billowing, and soaring — from almost anywhere in the surrounding hills.

Tom McAlister, in love with the beauty and accessibility of kites, founded the festival in a desire to give back to his community. "At the time," he points out, "most festivals were primarily either children's events or exhibitions for professionals. We wanted to offer all of the best that modern kiting had to offer."

McAlister, whose stepfather built gyrocopters, grew up loving anything that flew. He discovered kites in college, and started selling them out of the back of his Honda Civic in 1985. "You can be a kid or an 80-year-old in a wheelchair," he says, "and for a modest investment be cutting up the sky like the Red Baron."

Lucky for him, he was in kite-flying heaven: Berkeley, just opposite the Golden Gate, is directly in the path of some truly gorgeous and reliable wind. Over time, the weekend gig turned into a full-time job, and the festival blossomed, too. The timing was perfect: a park was being constructed on top of landfill when the festival began, and as the park grew, the festival swelled to fill it, now attracting crowds of up to 25,000 people every summer.

And diverse crowds they are. Your first sight once you actually get to the festival is a hilltop covered in picnic blankets and tangled kite strings, as little kids run amok trailing kites behind them. The variety is beautiful to see: there are elaborate, store-bought dragons clashing joyfully with colorful, homemade box kites and simple, hand-painted classic diamonds. The ecstasy is palpable. In the afternoon, everyone streams over to the demonstration area, where a kite showers down candy, tiding kids over until Halloween. "We had the first candy drop anywhere in the world," McAlister delights in mentioning. "It's kind of silly, but hey — we're proud of it."

He's just as proud of the festival's other acts. Over the hill from the families are the professionals: champion kite fliers mix with giant three-dimensional kittens dancing in the wind, team kite demonstrations share airspace with exhibition kites, and kite manufacturers like Prism, Revolution Kites, and Ozone let kite buffs test out their latest designs.

The bronzed Ray Bethell, a seemingly ageless multiple-kite-flying world champion, draws gasps from even the most seasoned kite watchers with his nimble fingers and whizzing stacks of kites. Kite aerial ballet groups perform with 96 kites at once, and 20,000 square feet of Peter Lynn's giant creature kites gallop, slither, and leap in the wind.

All in all, to say the event is "fun for the whole family" is understating it, but there is more in the appeal of kites than mere child's play. They have long been a symbol of freedom and hope (check out national newspaper archives for images of kite flying, and you'll often see articles about emerging democracy), and it seems hard to find someone who doesn't like them. McAlister, who has been designing his own kites for 15 years, is a firm believer in kites as an art form: "That doesn't mean every kite is art — and most aren't — but there's the potential, whether it's the design or the act of flying." When asked why he thought the festival had been so successful, he says: "A wise old man once told me, 'Kites are holding the wind in your hand.'"

Berkeley Kite Festival, July 29–30, 2006
www.highlinekites.com/Berkeley_Kite_Festival

Arwen O'Reilly is staff editor at MAKE.

Photography courtesy Highline Kites

Peter Lynn, who's been designing kites in New Zealand for well over 30 years, came up with his huge and magical trilobyte designs in the mid-1980s, and snagged the world record for largest kite in 1995. His creature kites (top) now represent all parts of the animal kingdom.

Ready for takeoff, dual-line stunt kite trains make a colorful ground display (left).

What Steam-Up would be complete without steam-powered personal transporters?

Boiler Room

A visit to the annual Yankee Steam-Up.

By Brian Jepson

Dozens of steam engines meet in Rhode Island every year, connect to a massive boiler, and crank away for all to see. Everything from the smallest tabletop engines to large industrial models share the steam and show off what they can do with it. This is the annual Yankee Steam-Up, held on the grounds of the New England Wireless and Steam Museum (users.ids.net/~newsm), an organization dedicated to preserving the beginnings of wireless and steam technologies.

Throughout the day, volunteers kept feeding the boiler as the steam engines kept cranking. And there are a few steam-powered vehicles prowling around as well: a steam-powered motorcycle and a vintage Stanley Steamer Bus were on hand.

In addition to this impressive display of steam power, the museum operates the oldest working wireless station in the world, Massie Wireless Station PJ. Because the spark gap transmitter's bandwidth is so wide, it can't be connected to an antenna without creating interference that would draw in the FCC's vans and black helicopters. The museum has a collection of Morse code keys, radios, television equipment, and transmitters — all vintage and antique equipment from throughout the history of wireless communications.

Whether handmade or rescued and restored from an earlier time, all of the engines and wireless equipment have felt a maker's touch. Museum members meet every Thursday to tackle the work-in-progress. The Yankee Steam-Up is the museum's major public event. To see the museum at other times, visitors should contact the museum in advance at (401) 885-0545.

Brian Jepson is an O'Reilly editor, programmer, and co-author of three O'Reilly Media books.

Photograph by Brian Jepson

Ship of Cards

Popular card games become an exercise in miniature construction. By Alex Handy

THE WORLD OF COLLECTIBLE CARD games has long been dominated by Magic: The Gathering and Yu Gi Oh. But the 2D nature of these games has recently been tossed aside and the world of card games has a new caveat: *to play, you must first build*.

Up-and-coming games company WizKids, founded by Jordan Weissman, created Pirates of the Spanish Main in 2004. It's a collectible *constructible* card game based around the idea of using desks and tables as a substitute for a board, and tiny punch-out boats as the pieces in a game of high-seas combat and plunder.

Since then, WizKids and its competitors have taken this concept to new extremes, with RocketMen, a 1950s serial-style space game; a NASCAR-based racing game; a driving/jousting game called Racer Knights; and the catchall game, Z Cardz.

No matter who makes them, constructible card games all have one thing in common: you have to build stuff to play. Game pieces come in the form of perforated polystyrene cards, the parts punched out and assembled into 3D in-game characters, ships, and weapons. First came pirate ships; now there are pterodactyls, fleets of spaceships, and even race cars that roll like their Matchbox equivalents.

And it all started with Mike Mulvihill, a member of the game design team at WizKids and a grizzled veteran of the game design world. "There was such a resurgence in pirates, but we couldn't afford to make ship miniatures," he says.

"When the styrene thing fell into our laps, we knew we could do ships with the plastic. But then the question was, if you do this, how do you make everything in that pack of cards become something that is useful to play the game with? It was like Apollo 13."

Mulvihill and his team eventually built Pirates of the Spanish Main into a game that can be played by two people with only two packs of cards. Total price to play is about $8. Each game pack comes with a tiny die, a punch-out island, two ships, a card filled with gold punch-out pieces, and a card full of crew.

Since playing the game typically requires more than two islands, the instructions suggest using the frame left over from the island card as an island itself. More islands can be made out of the leftover pieces of ships, and even the card pack wrappers. During a game, ships move set distances that are measured by lines drawn along the edges of the empty cards that once housed masts, hulls, and deck plates. The game is truly a wonder of resourcefulness.

Alex Handy is a freelance journogeek, and his office is now adorned with little styrene pirate ships. He blogs at gism.net.

Photograph by Alex Handy

Space Cases

The balloon men at NASA's Jet Propulsion Laboratory.

By David Pescovitz

In 1835, an article appeared in the *Southern Literary Messenger* telling the remarkable story of one Hans Pfaall, a bellows-mender from Rotterdam who escaped his creditors by flying a homemade hot air balloon to the moon. The story turned out to be a prototypical bit of science fiction dreamt up by a young Edgar Allen Poe to dupe his readers. The idea of ballooning to the moon or elsewhere in the solar system never took off, probably because it's impossible. But for a certain group of engineers, flying balloons on other planets isn't far-fetched at all. In fact, the Russians have already done it. And NASA's Jet Propulsion Laboratory (JPL) is keeping the tradition alive.

A sprawling campus just north of Los Angeles, NASA's Jet Propulsion Laboratory was established by the California Institute of Technology in the 1930s. Emerging from the DIY mindset that launched the science of rocketry in this country, JPL was where the first U.S. satellite was designed, the Mariner and Voyager probes were born, and the long-running Mars rovers were built.

In a small cluster of bunkers at the heart of the campus lies JPL's Mobility and Robotic Systems laboratory. About 100 engineers develop all of the robotic components, from the autonomy software to the mechanical appendages that enable us to explore other worlds while staying safely on terra firma.

In a back room of the laboratory, a white blimp, about 15 feet long, floats above a table, secured by tethers. The blimp doesn't seem particularly unique, that is until you hear that it's a prototype for an autonomous nuclear-powered aerobot that could someday explore Saturn's moon Titan. The blimp would fly below the dense clouds that hide the terrain from orbiting spacecraft, snapping high-res photos and possibly even scooping up surface samples for onboard analysis.

"Balloons provide a unique observation platform for doing planetary science that you simply can't do any other way," says Jeffery Hall, the senior engineer who leads JPL's aerobot research.

An aeronautical engineer by training, Hall joined JPL in 1997 after graduating from CalTech down the road. At the time, he worked on cryogenic technologies for satellites. Then, he says, the funding dried up and he was forced to expand his horizons.

"I've always been motivated to figure out how things work, and that inevitably leads you to building stuff for experiments," he says. "One of the great attractions of balloons is that it's fairly easy to make them. Then you get to take them outside and fly them around. From an experimentalist's point of view, they're fun things to work with."

The Titan balloon is just one of several very different aerial vehicles that Hall and his colleagues are designing. Each vehicle's technology is dictated by its ultimate destination. For example, the very thin atmosphere of Mars requires a spherical balloon at least ten meters in diameter to carry just a couple kilograms of scientific instruments.

In one scheme, the balloon package would be released from a spacecraft upon entry into the Mars atmosphere. As the canopy drifts down beneath a parachute, onboard helium tanks inflate the "envelope," the actual balloon fabric. Once the envelope is filled, the parachute and tanks are cut loose and the balloon settles at an altitude a few miles above the surface. According to Hall, a wind-blown helium superpressure balloon like this could float around the planet for up to a year, all the while transmitting data back to mission control on Earth.

Photography by Chris Leschinsky

Viktor Kerzhanovich stands beside the JPL prototype aerobot. During high-altitude flight tests, internal combustion engines provide enough propeller power to overcome Earth winds.

Following page: On Titan, the juice would come from an onboard nuclear source that would "effectively last forever," says Jeffery Hall (pictured).

To put the technology through its paces, the researchers have flown their balloons in the environment closest to the Martian atmosphere that's readily accessible — well, relatively accessible anyway: 100,000 feet up in Earth's stratosphere.

"It's a difficult place to do experiments, but that's where you have to go if you want something similar to what you'd see on Mars," Hall says.

A commercial scientific balloon tows the rolled-up test balloons to the right altitude. From the ground, the researchers activate a pyrotechnic cutter that releases the packaged prototype with accompanying helium tanks. Then, they cross their fingers and watch the live video feed.

"We've had a lot of failures and a couple of successes under our belt," Hall says.

The Jet Propulsion Laboratory didn't invent the idea of flying balloons on other planets. Early in the summer of 1985, a pair of superpressure helium balloons floated through the upper atmosphere of Venus. They were part of the Soviet Union's Vega mission to study the planet's atmosphere and surface.

The windblown balloons, 12 feet in diameter and sewn from a Teflon-like fabric, carried a gondola outfitted with thermometers, velocity sensors, light and pressure sensors, and a nephelometer to measure cloud particles. The high winds quickly swept the balloons across the dark side of the planet at an altitude of 35 miles while their progress was tracked by radio telescopes on Earth. After two days, the batteries died as expected, contact was lost, and the balloon mission was declared a resounding success.

The JPL Robotics researchers have inside knowledge of the Vega mission. That's because one of the lead engineers on that project is now on their team. As part of his lifelong career in the Russian space program, Viktor Kerzhanovich worked on 15 of the USSR's Mars and Venera missions, including Vega.

"In the Soviet Union of the 1980s, the space program tried several noble things and the balloons were one of them," he says.

While a follow-up balloon mission to Mars was planned by the USSR and the French Centre National d'Etudes Spatiales (CNES), the collapse of the Soviet Union essentially wiped out Russia's pioneering space science program before the technical challenges could be met. Kerzhanovich knew it was time to take his talents elsewhere.

In 1994, Kerzhanovich emigrated to California where he worked briefly at an aerospace engineering consultancy before joining JPL. While he felt fortunate to have the opportunity to follow his passion, the difference in culture was like, well, visiting another planet.

"The (space program) here is much more bureaucratic than it was in Russia back then," he says. "There, the governmental role was not great. Most of the decisions were made by the projects' chief designers. Here everything goes from the top to the bottom. The process is much more regulated and it's directed to avoid as much risk as possible."

Kerzhanovich may once again have the chance to fly balloons on Venus. He and Hall are applying to NASA's Discovery Program, an effort to launch many small missions with short development cycles, for a flight mission to Venus. (JPL's Stardust mission that brought comet samples back to Earth was the most recent Discovery Program mission.) The Venus trip would be a long-awaited encore of the pioneering Russian mission, but the scientific payload would be far more elaborate, including a mass spectrometer to analyze the planet's atmosphere.

The balloon material is also much more advanced. Developed by JPL with private-sector collaboration, the fabric is built to withstand the Venusian clouds of sulfuric acid. It's a multilayer laminate material consisting of Mylar, Teflon, and Vectran, the high-strength fabric used for the airbags that cushioned the fall of the Mars rovers. Right now, engineers are constructing an 18-foot spherical balloon from the material to help sell the proposal.

"It's not common to make prototypes as part of these proposals," Hall says. "But JPL ponied up money to make one so we can demonstrate the technical maturity of this technology and convince skeptical reviewers. After all, they'll get 15 or 20 proposals submitted for just one or two flight missions. Not everyone gets to go."

And if this particular design doesn't fly, there will certainly be other shots. After all, the space race may have ended, but the makers at JPL still have their eyes on the sky.

"I don't know if I could work like I do if I didn't think one of these projects would fly in the relatively near future," Kerzhanovich says.

David Pescovitz, MAKE's editor at large, is also co-editor of BoingBoing.net and a research affiliate with the Institute for the Future.

The blimp doesn't seem particularly unique, that is until you hear that it's a prototype for an autonomous nuclear-powered aerobot that could someday explore Saturn's moon Titan.

Distributist Technologies

At last, Tolkien meets capitalism.

By Tom Owad

IN *THE LORD OF THE RINGS*, J.R.R. TOLKIEN contrasts the bucolic "hobbit sense" of the Shire with the noxious industrialism of Mordor: "The one small garden of a free gardener was all his need and due, not a garden swollen to a realm; his own hands to use it, not the hands of others to command."

Communal in structure, nonmechanical, and inhabited by individual craftsmen and farmers, the Shire clearly parallels preindustrial England. Its ravaging bears an epic similarity to the transformations that have taken place in our own world since the Industrial Revolution.

As was abundantly clear in Tolkien's time, to control the means of production is to control life. Industrial capitalism placed this control in the hands of a wealthy elite; communism placed it with the state. Whether at the Carnegie Steel Company or the Lenin Steelworks, and whether the bullets came from the Pinkertons or the Red Army, the power over life and death did not rest with the individual but with a hostile external force.

The distributists of the early 20th century rejected both visions. "Too much capitalism," wrote G.K. Chesterton, "does not mean too many capitalists, but too few." Drawing upon the principles of subsidiarity and solidarity put forth by Pope Leo XIII's social encyclical *Rerum Novarum*, the distributists sought a society where each individual provided both his own labor and his own capital. The small garden of Sam Gamgee was the distributist ideal.

Distributist principles took root all over the world, but the luxuries of the Industrial Revolution triumphed over the freedom of the homestead. A similar phenomenon occurred in the 1970s, with the introduction of Schumacher's economic treatise *Small Is Beautiful* and the back-to-the-land movement. Urban idealists also found themselves ill-equipped and unprepared to deal with the hardships of living an essentially preindustrial agrarian life.

Today, freedom and technology no longer stand in juxtaposition. A CNC lathe or milling machine can be built for $500. Raw materials can be prepared in homemade foundries. Solar power and wind turbines make it possible to live comfortably off the grid.

Open source software is ubiquitous, and hardware is following the same path. The pages of MAKE illustrate the countless sophisticated devices that can be built without a factory. The internet allows for worldwide collaboration on a level never before possible. Goods can be exchanged between individual craftsmen without the need for distribution networks and middlemen.

Advancements in personal fabrication at the MIT Center for Bits and Atoms hold the greatest promise. Their aim is to produce a personal fabricator that can manufacture anything from a doll to a precise replica of itself. The current MIT Fab Lab (*see MAKE, Volume 01, page 23*), which has been deployed in nine countries, is itself an impressive feat. The equipment can be assembled for $20,000; the software is open source and free. For the price of a new car, it's already possible to establish an effective personal fabrication laboratory.

Tools such as these have the potential to personalize the Industrial Revolution — to place the means of production not in the control of select capitalists or of the state, but with every individual seeking to partake in the act of creation. Like the soil, the machine yields in response to the labor of our hands.

Tom Owad (owad@applefritter.com) is a Macintosh consultant in York, Pa., and editor of applefritter.com.

Illustration by eboy.com

RG 259 9
PL 269 9
PR 279 9
VISA
MasterCard

Don't assume he's wearing a pair of old-fashioned *pince-nez* eyeglasses. Look more closely and you'll see there's nothing pinching the bridge of his nose. In fact, they're attached by magnets on a steel rod that pierces the fleshy part above the bridge of his nose.

You Are The Platform

How hardware hackers are remaking their bodies.

By Quinn Norton

THE TERM *BODY MODIFICATION* BRINGS to mind piercing and tattooing for cosmetic reasons. But body mods can also be assistive, by changing a person's experience of their own body or even giving them a new ability. How would you like to be able to sense an electric field at a distance? Talk to Steve Haworth of Phoenix (stevehaworth.com). He has experimented on human volunteers by making a small incision in the fingertip and inserting a small neodyme magnet coated in gold, then silicone. The magnet is nestled into a bed of nerves between the epidermis and fascia layers of the ring finger, and the cut is neatly sutured up. For most magnetic implants, Haworth can have the job done in less than ten minutes.

Over the next few weeks, the finger heals, and the nerves begin to interpret electromagnetism and movement in the magnet. Some things, like other magnets, can even cause the magnet to spin freely up against the nerves. The new sensations pass to the brain, where evolution's own favorite primate hack takes over — parts of the brain become more in tune with the new signals from the hand, and the perceptual sensitivity improves. How much sensitivity

Photograph courtesy of James Sooy

you experience depends on many factors — the size and placement of the magnets, how many you have, and, probably, how well you train to use them.

Every magnetically modded person tested, from two weeks to several months, could easily pick out a live cord running with 110 volts. Many report being able to sense electrical motors running from distances of a few inches to a few feet, depending on the strength of the motor and the size and healing time of the implant.

Shannon Larratt, of the body modding publication and community BME, reported that walking through a retail security device was like "sticking my hand in an ultrasonic cleaner." Any strong inductive wire has a chance of causing a sensation in the magnetic finger.

Every magnetically modded person tested could easily pick out a live cord running with 110 volts.

(For a time, Haworth stopped implanting the magnets because, in some cases, the medical-grade silicone sheathing had broken, exposing the magnets' metal to the body. He is now using a harder type of silicone to make the implant safer.)

LOOK MA, NO BRIDGE!

When you meet James Sooy, you might initially assume he's wearing a pair of old-fashioned *pince-nez* eyeglasses. But if you look more closely, you'll notice there's nothing pinching the bridge of his nose. Each lens seems to be glued to either side of his nose. In fact, they're attached by magnets on a rod of steel that pierces the fleshy part above the bridge of his nose. They're more like *pierce-nez* glasses (piercedglasses.com).

Cheaper and less permanent than Lasik, easier to cope with than contacts, Sooy chose piercing to get rid of the earpieces of his glasses. While he plans to make a product, for now it's strictly DIY.

If you want your own pair, you'll need: a small tabletop mill, four small rare earth magnets, a bridge barbell piercing, six small screws that fit the barbell piercing, and a pair of plastic eyeglass lenses.

The first step to pierced glasses is the bridge piercing itself: a barbell that goes through the skin above the bridge of the nose. Get internally threaded jewelry — this is a good idea in general, and very important in this case. "I'd suggest to wait at least a couple months, healing time, before playing around with any piercing," says Sooy. Next, you're going to need screws that match your barbell. "We called up the piercing studio to find out exactly what threading the barbell was, and just ordered the screws."

The elbow brackets that fit on the end of the barbell were cut out of aluminum on a tabletop mill. There's some artistic license here, but make sure you leave a front-facing divot to fit in a set of rare earth magnets. It's the magnets that actually hold on the glasses — and make it safe for them to be knocked off your face. Hold the lenses up to your piercing and mark the drilling spot to mount the fitting for the other pair of magnets. Don't discard the nose pads — as with traditional glasses, it's actually the nose pads that hold the weight.

Quinn Norton is a freelance writer and co-blogger at ambiguous.org.

Health and Safety

Your body modification procedure will involve both pain and blood. The magnetic implant pain is worse than a simple injection or having your ears pierced, but not as bad as a long tattoo over a sensitive region, or many medical procedures.

For any piercing or modification, find a local professional who follows careful protocols for dealing with biohazardous waste — which is what blood is. Ask to see their autoclave, and ask how often they calibrate it; if they are reluctant to answer, leave. More info is available from the Association of Professional Piercers (safepiercing.org).

Any body modification, functional, aesthetic, sensual, or fun, needs to be carefully healed. This is an open wound, and needs to be properly cleaned and sutured, where appropriate. Make sure you get aftercare instructions and understand them.

During the healing period, don't be overly rough on the modified area or expose it to unclean conditions. Ask how long the healing period normally takes for your type of piercing. Keep an eye out for signs of infection: fever, pus discharge, excessive swelling, or redness, and talk to your modder or doctor if you notice any of these.

Saul Griffith

SIMPLY CAD

COMPUTER-AIDED DESIGN DOESN'T HAVE TO BE AWFUL. THERE'S A CAD PROGRAM OUT THERE FOR YOU.

HAVING USED VIRTUALLY EVERY CAD (computer-aided design) program out there at least once, I thought there was no chance they would ever be simple ... until I met Takeo Igarashi. Takeo was a grad student from Japan who wrote an incredible little piece of software called Teddy. It's basically CAD for 5-year-olds: if you want to design a plush toy, this is the program. It's intuitive. It's beautiful. CAD doesn't have to be awful or expensive. It doesn't have to come with 1,000-page instruction manuals.

My hope was that Teddy would influence the world of CAD and make incredible packages cheap and easy to use. That was six years ago. Things haven't changed that much, but there is still hope.

Why, as a reader of MAKE, should you care? I care because I just enjoy using CAD; it's better than playing a computer game. You can spend the same 36 sleepless hours in front of the computer, but at the end of it, you have a beautiful 3D object from your imagination, planned out and ready to be built or shared. But the real and exciting reason to care is because someday kids should be able to unleash their imaginations on a Teddy-type CAD program and then print their imaginings into the real world on a chocolate- or sugar-powder 3D printer.

You should care because the capacity to share 2D and 3D models is at least as important as sharing music and video. Making CAD representations of your "makings" makes it easier for other makers to make things too.

Before I launch into my CAD diatribe, let me sketch out the playing field for the uninitiated. Very broadly, CAD programs can be classed in a number of ways. On the hardcore engineering side, there is parametric 3D solid modeling, with exact equation-driven descriptions of objects; there are surfacing or wire-frame-based programs, which bring nice rounded surfaces and squishy shapes to life; there's a world of rendering tools for making the 3D models look cool; and there are architectural-type 2D drafting programs for laying out plans in 2D.

For completeness, I should also mention CAM, which is computer-aided manufacturing software. It takes your CAD design and converts it into a "tool-path" for a drillbit, mill, laser, inkjet head, or some other tool to follow and construct your object.

Contrary to popular belief, you don't need a laser cutter or a 3D printer or an NC mill to realize your CAD designs. On your desk, you likely already have a high-precision machine tool: an inkjet printer. These things will print designs onto paper with 50-micron,

> Build your dream house and stick it in Google Earth! How cool is that?

or better, accuracy. I very often use the inkjet printer to print a pattern or template, which then can be used to align the holes you will drill by hand or to cut out the fabric you will stitch together. So, in fact, the world already has the capacity to build cool CAD 3D models and share them via their desktops.

One program taking advantage of this is Pepakura Designer — it calculates a pattern of triangles that, when folded up, can make your 3D object. It's high-tech origami. I've seen business cards that fold into toy cars made by this technique, and Japanese anime characters and dinosaurs.

In the same vein is a program called Lamina Design, which takes 3D curved surfaces and figures out how to construct them from flat sheets. Again, the nice thing here is that you don't need NC machine tools to get real-world output; you could use this program, print to paper, trace to plywood, cut with a hacksaw, and make beautiful organic furniture, or even design your own kayaks.

A great little company called SketchUp has a CAD package that allows you to build buildings and place the 3D models in Google Earth. Of all the packages

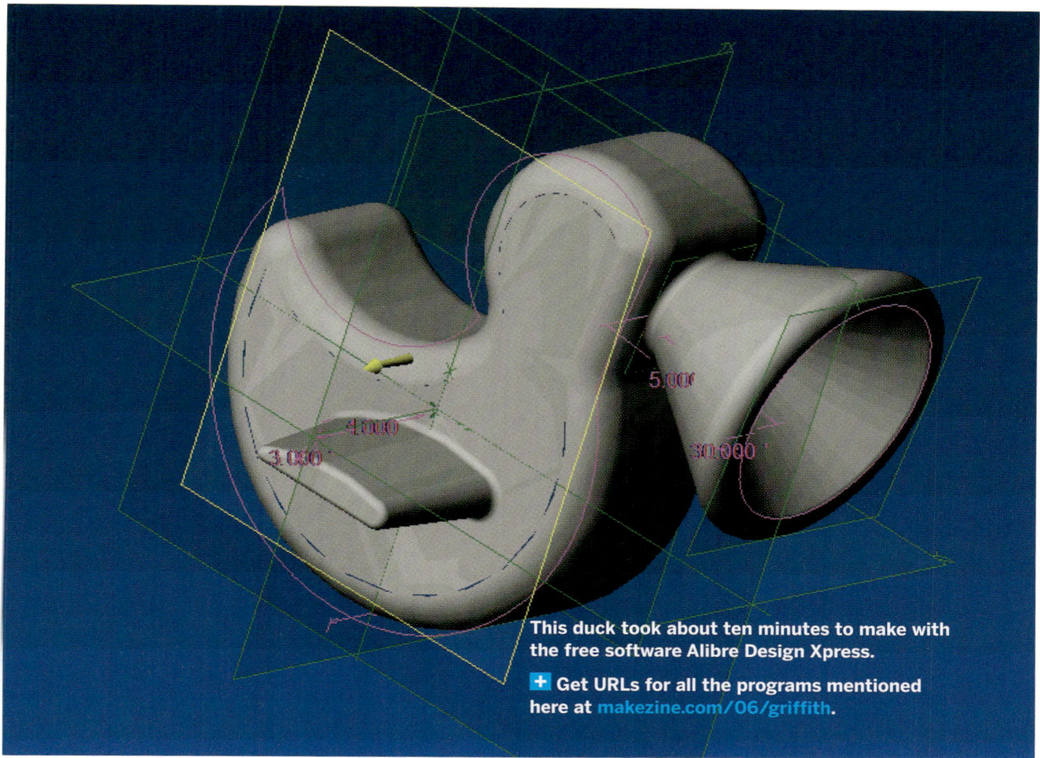

This duck took about ten minutes to make with the free software Alibre Design Xpress.

+ Get URLs for all the programs mentioned here at makezine.com/06/griffith.

out there, SketchUp is the closest in my mind to the Teddy ideal of simplicity. They have great tutorials and an intuitive interface, and if you've never used CAD before, you'll just enjoy this. Build your dream house and stick it in Google Earth! How cool is that?

The online world Second Life now has CAD packages built into it, so you can realize fantasy objects in the other world. This is cool to me because I can now see real/virtual world crossover. Make something über-cool in Second Life? Bring it into the first life!

So where should you get started? Before you spend a few thousand dollars on a pro CAD package, try some free alternatives: freeCAD.com reviews a bunch. My favorite right now is Alibre Design Xpress. It's a free parametric CAD program with most of the functionality of the pro packages.

Their business model is interesting (a professor of mine once described it as "crack for babies"): they give you the free software, you get addicted, and then if you need higher functionality (more parts, more tools), you can purchase more. Their prices are reasonable, and you only purchase the extras you need.

If you're comfortable in Illustrator or CorelDRAW, you can do a surprising amount in these programs. There are dimensioning plug-ins for Illustrator, and with these programs you can produce nice 2D plans of your wooden boat, air-powered potato cannon, or cigar box guitar. Eagle CAD is a free circuit board design and layout tool. If electronics is your gig, you can CAD it up in Eagle and split the cost of sending out for boards with your friends who also want mint-box MP3 players.

Another idea is to write your own CAD program. David Aberdeen wrote a fantastic CAD package for kite design, called Surfplan, that led a whole community of amateur kite builders to build cool kites for surfing. A lot of people have built small CAD packages within MATLAB and Mathematica with outputs for DXF or PostScript. Robert Lang uses such a thing to create his awesome computational origami models.

The world of CAD still isn't perfect. Converting between platforms is a nightmare. Learning curves are long. Things have come a long way, though, and there are a few packages to get you started with little pain. Hopefully, out there in the MAKE readership, there are hackers who will write cool new CAD programs with simple interfaces and open source extensibility. Get to work.

Saul Griffith thinks about open source hardware while working with the powernerds at Squid Labs (squid-labs.com).

CAD image by Saul Griffith

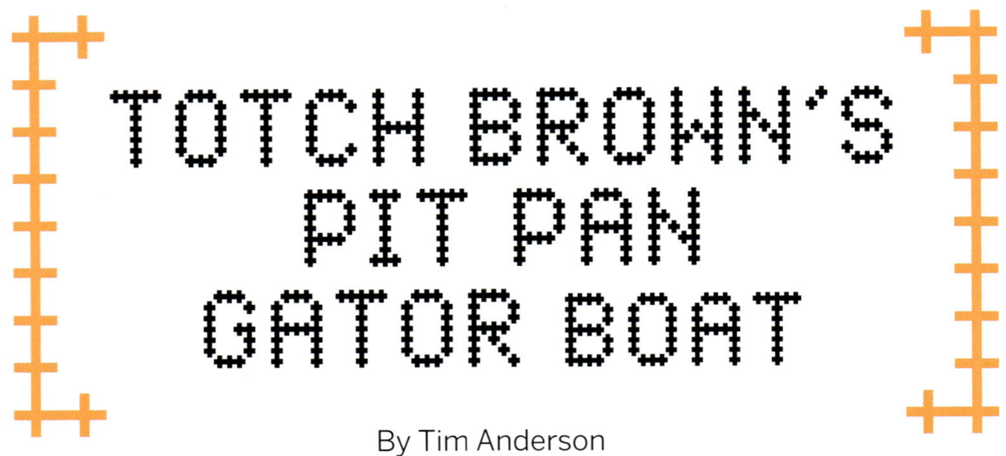

TOTCH BROWN'S PIT PAN GATOR BOAT

By Tim Anderson

START YOUR BOAT-BUILDING HOBBY BY BUILDING THIS ONE.

Loren "Totch" Brown lived in the Thousand Islands region of the Everglades his whole life. His fascinating autobiography, *Totch, A Life in the Everglades* (University of Florida Press, 1993), relates his adventures, and its cover depicts him propelling a strange little boat with a push pole.

It's called a "pit pan" and is now on display at Smallwood's store in Chokoloskee, Fla. (florida-everglades.com/chokol/smallw.htm).

When the water was too deep for the push pole, he used a paddle, and when the water was too shallow, he dragged or carried it.

He made many such boats for hunting alligators over the years. Despite its crude appearance, this is a very elegant boat. It is burdened with no unnecessary features. You can carry it and you can sleep in it. That makes it a magic carpet to freedom.

In designing boats, people get hung up about the speed of the boat rather than the speed of building the boat, and they forget that to go fast you have to work hard. If you want a boat you can take a nap in, you need something like this.

Rolled up in the bottom of the boat is Totch's mosquito net or "skeeter bar," so named because it's where the insects go to drink and socialize. It's a rectangular mosquito net just big enough to not touch the person sleeping under it. In the old days, gauze, cheesecloth, or any open weave fabric was used. The net is held up at the corners by sticks. If it rains, a tarp just big enough to keep the rain off goes over that.

A boat can be part swan, part guitar. This boat is a box with no lid. Start your boat-building hobby by building this one.

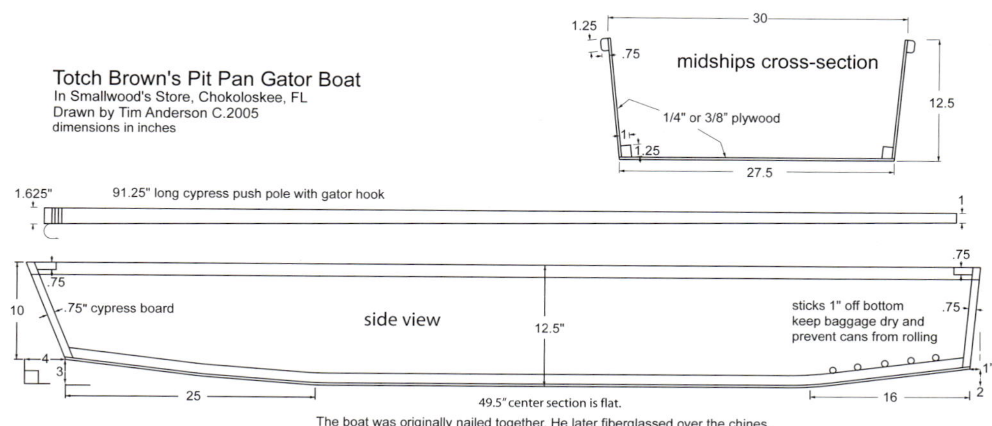

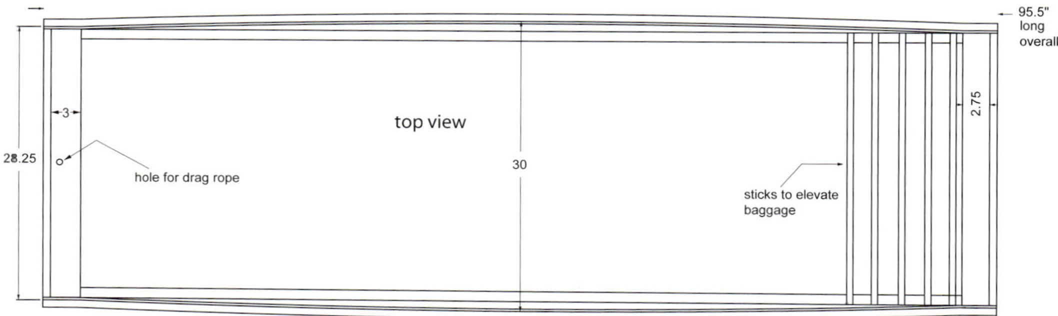

The plans for the pit pan.

Attach all the pieces together with nails and screws. To make the boat watertight, paint the mating surfaces of each joint with epoxy glue before nailing. To keep the epoxy from running out of the joints, add enough white flour to thicken it to the consistency of pancake batter. Fine wood dust from sanding works also. Whole wheat flour makes lumps and isn't as good as white flour for this. Don't use 5-minute epoxy for anything; it's not fully water resistant. To get more working time with your glue, pour it out onto a pan. That will keep it from getting too hot and curing before you're ready.

Paint the whole boat with linseed oil, and after a couple of days, paint the whole boat with whatever paint you can find, preferably oil-based. Roofing tar is also really good for sealing the joints.

Totch covered the outside edges of his boat with fiberglass, which is also a good method.

If you don't have plywood, Dan Beard's 1882 classic, *The American Boy's Handy Book* (see Figure 1), shows how to build this boat from boards. Beard, the founder of the Boy Scouts of America, suggests putting paint-soaked strips of wool between the boards to seal them.

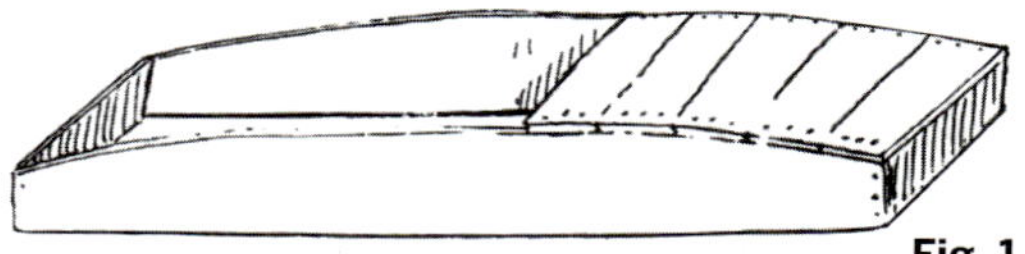

Fig. 1

EVERGLADES AND SWAMP BOAT RESOURCES

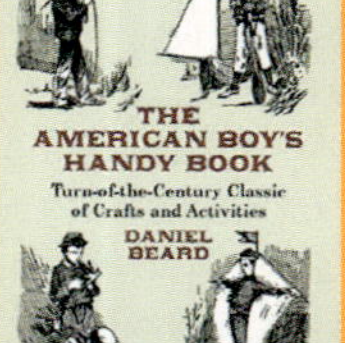

Totch died in 1996, but his website is still online at florida-everglades.com/totch/totch.htm

You can order a facsimile edition of *The American Boy's Handy Book: Turn-of-the-Century Classic of Crafts and Activities* from Dover at store.doverpublications.com/048643138x.html

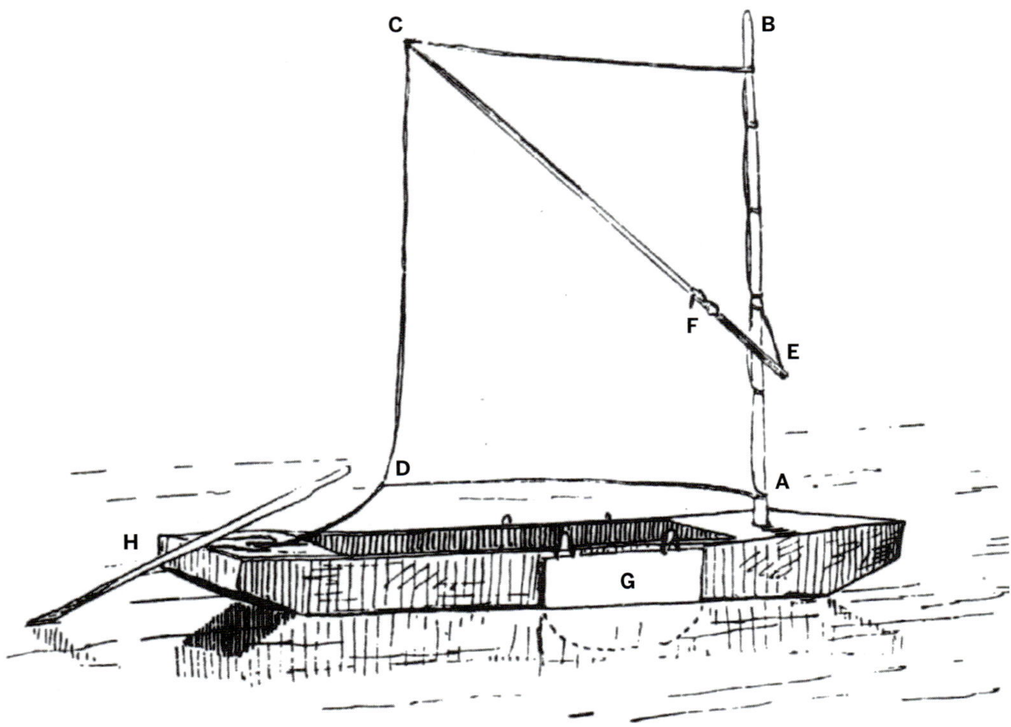

Fig. 2

SHOVING OFF

Want to try sailing your new boat? Figure 2 from the same book shows a good spritsail rig for this boat. The push pole becomes a mast, and the tarp becomes a sail.

A leeboard (G) hangs over each side to keep the boat from drifting sideways. (A "no parking" sign works well for this.) The sail is a rectangular tarp, sheet, poncho, or tablecloth tied to the mast and sprit at the corners. The mast (B) and sprit (C) are short enough to stow in the boat. The mast can be stepped off-center, if you prefer. The sprit has a notch (E) at its end for the "snotter line." The snotter is tightened at the cleat (F) to adjust the sail. In light winds the sail doesn't need to be fully laced to the mast. Just the corners is fine.

Figure 3 shows a simpler way of attaching the sprit. The snotter is a loop of cord formed into "cow hitches" on the mast and sprit. The sail is "peaked" or adjusted by sliding the cow hitches up and down. To stow the sail, lift the notched end of the sprit off the snotter. That will give you enough slack to put the mast and sprit together, so you can roll the sail up on them.

Fig. 3

Attaching the sprit to the mast: The cord attached to the sprit (diagonal) and mast (vertical) is called a "snotter," with "cow hitches" on either end.

Fig. 4

If you prefer rowing to paddling, these oarlocks are easy to build. Any scrap wood will do, but they take a fair amount of friction and pressure, so the wood should be strong. Make sure to anchor them securely.

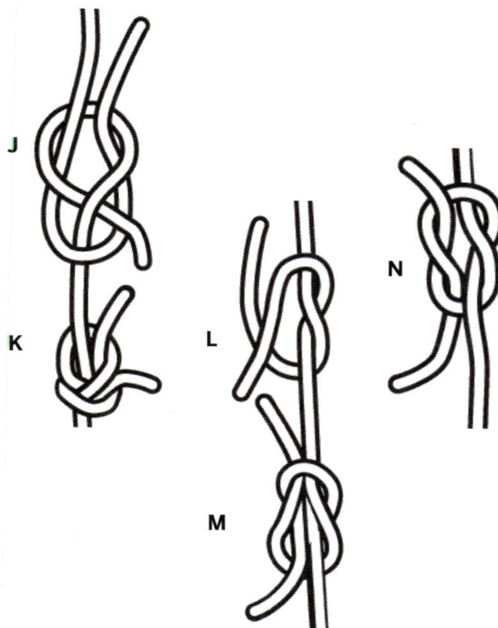

KNOTS

Useful knots for the sailor begin with the "sheet bend," a.k.a. "bowline" (J and K). Learn it well and don't use any other knot without a good reason. This knot will not jam up. It can join dissimilar ropes.

The square knot, a.k.a. "reef knot" (L and M) is easy to untie, which can be a good thing. It only works with same-size rope and must be properly formed to hold.

The "granny knot" (N) is often tied by mistake instead of the square knot. This knot slips, causing many tragedies. I'd always been told that the granny knot was only ever a mistake and sometimes a tragic one. Then an Ojibway female canoe builder showed me how to tie split spruce root with a granny knot. Other knots bend the material too much and it may break. So even the granny knot has a use, and a pretty good one at that.

For more knots, see realknots.com/knots.

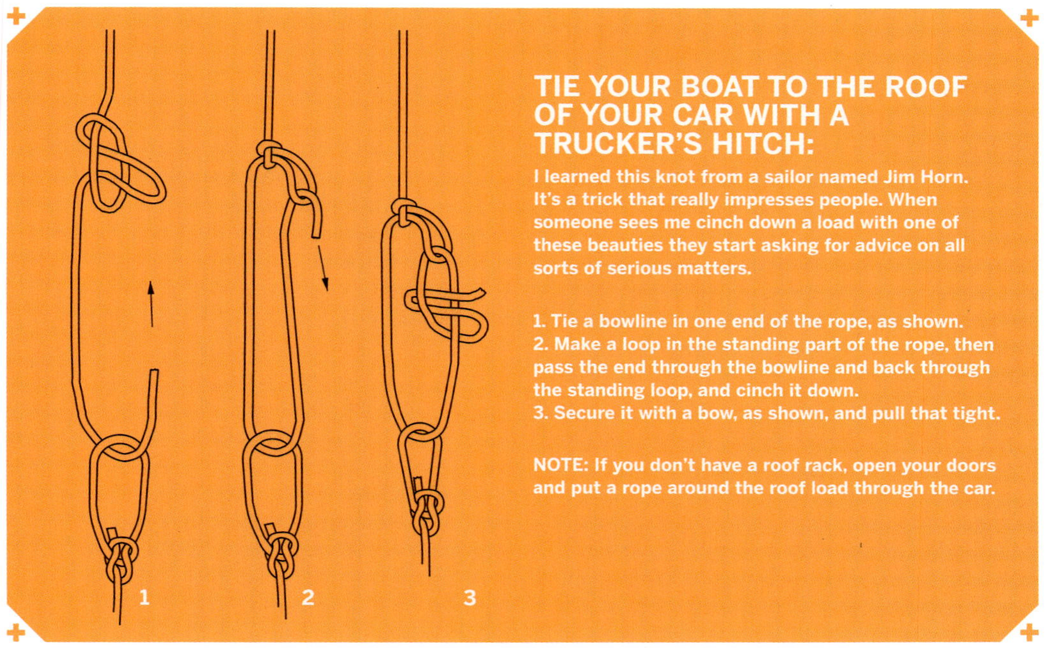

TIE YOUR BOAT TO THE ROOF OF YOUR CAR WITH A TRUCKER'S HITCH:

I learned this knot from a sailor named Jim Horn. It's a trick that really impresses people. When someone sees me cinch down a load with one of these beauties they start asking for advice on all sorts of serious matters.

1. Tie a bowline in one end of the rope, as shown.
2. Make a loop in the standing part of the rope, then pass the end through the bowline and back through the standing loop, and cinch it down.
3. Secure it with a bow, as shown, and pull that tight.

NOTE: If you don't have a roof rack, open your doors and put a rope around the roof load through the car.

Illustrations by Tim Anderson and excerpted from *The American Boy's Handy Book* 1882

Tim Anderson, founder of Z Corp., has a home at mit.edu/robot.

A Beginner's Guide to BEAM

The BEAM design approach creates nimble robots from simple components, with no programming required.

By Gareth Branwyn

EXPECTATIONS FOR CREATING THINKING and actuating machines were high in the 1960s and 70s, as universities began establishing robotics and artificial intelligence labs. But researchers soon realized that even the most basic physical tasks, such as getting a cart to sense obstacles in a space, plan a route, and execute it (called *sense-plan-act* architecture) were daunting assignments. Some wrestled with these problems by developing tweakier algorithms and throwing more, faster hardware at the problems, but MIT's Rodney Brooks took a radically different tack. Inspired by insects and other teeny-brained critters, he wondered what a *sense-act* architecture might look like, one where the bots didn't bother to create a map of their world to plan from, but rather, reacted directly to environmental stimuli.

The results were impressive: bug-like bots that could do most of what the sense-plan-act robots could do, but with scant little computing power. Many roboticists were inspired by Brooks' work, and one, Mark Tilden, took the idea even further. After seeing Brooks lecture in the early 90s, he wondered whether it was possible to create sense-act robots that used no digital computation at all. The answer was yes, and BEAM robotics was born.

The initials "B.E.A.M." stand for the four components of the BEAM design approach:

Biology: Like other sciences and technologies these days, BEAM looks to nature for design inspiration. After all, billions of years of design, fabrication, and in-field testing shouldn't be taken lightly. BEAM also encourages the human's role in robotic evolution. Olympic-style BEAM competitions are held each year, and in the spirit of hacking and open sourcing, builders share their winning design concepts for incorporation into future robotic competitors. In BEAM evolution, humans are the way for a robot to make a better robot.

Electronics: A hallmark of BEAM technology is the cyberpunky "abuse" of electronics, re-purposing components in ways that their designers never intended. Chips used in common household electronics to amplify sounds, direct data traffic, or invert signals are connected in new ways in order to control motors, process sensor input, and act as state timers. Chain some of these sub-circuits together, and you have a robot brain with no need for a programmable microcontroller.

Aesthetics: Buckminster Fuller once said: "When I'm working on a problem, I never think about beauty; I only think about how to solve the problem. But when I have finished, if the solution is not beautiful, I know it is wrong." Tilden and fellow BEAMers understand that elegance in robot design serves the evolutionary purpose (see Biology) of attracting the enthusiasm of fellow builders, which leads to replication. Many also take pains to make their creations beautiful because they see their robots as a form of kinetic sculptural art.

Mechanics: Like BEAM's electronics, its mechanics and build techniques are often quite clever in how they maximize efficiency. For instance, many BEAM bots use free-formed circuitry, where the components are soldered directly to each other rather than to a printed circuit board. This lowers cost and weight, and it shows off the engineering prowess and soldering skills of the builder.

Harold R. Ilano's Black Vermin 1.1, a replica of Mark Tilden's infamous Vbug 1.5, using hobby servos instead of Walkman motors. More images, schematics, and video at raysbeambots.solarbotics.net/harold/harold.htm

A few other key BEAM precepts are worth detailing, as they say a lot about how BEAM differs from other robot design approaches.

Keep It Simple, Stupid (K.I.S.S.)

BEAM's Zen-like simplicity strips away assumptions to get at the essence of a problem. Look at the Robosapien, a toy humanoid robot designed by Tilden that incorporates many BEAM concepts. Robosapien uses almost no digital processing, yet it performs many of the behaviors of corporate research robots like Sony Qrio and Honda ASIMO. A Robosapien costs less than $100, while these high-end industrial fussybots aren't even on the market — and would cost a fortune if they were.

> 1. Protect thy ass. 2. Feed thy ass. 3. Get thy ass to better real estate.

Junkbots and Appropriate Technology

BEAM also emphasizes recycling. One of Tilden's early BEAMbots, Vbug 1.5, was made from little more than a few cheap analog ICs, parts lifted from several Sony Pro Walkmans, an oven timer, and some wire. BEAM enthusiasts do use microcontrollers and other high-end components, but only if cheaper and simpler solutions are not possible.

My Many Dumb Bots Beat Your One Smart Bot

Implied within the BEAM philosophy is Brooks' idea that a bunch of working dumb bots is better than one broken smart bot. BEAM enthusiasts envision a future in which swarms of cheap robots could do such things as clean skyscrapers, toxic waste sites, and even your house. If a few of them break or run out of juice, that's fine, but if a single high-priced fussybot breaks so much as one end effector, you're out of luck.

Real Robots Fend for Themselves

Whenever possible, BEAM builders design bots that can take care of themselves. Many older designs are solar-powered. Some more recent BEAM and BEAM-inspired designs do require batteries, but ingenious engineering maximizes their efficiency.

Photograph by Harold R. Ilano/haroldsbeambugs.solarbotics.net

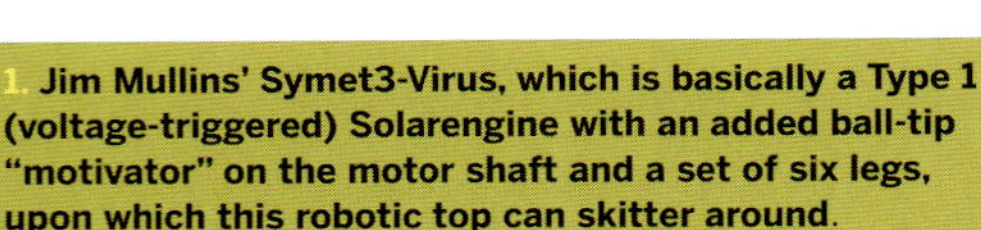

1. **Jim Mullins' Symet3-Virus, which is basically a Type 1 (voltage-triggered) Solarengine with an added ball-tip "motivator" on the motor shaft and a set of six legs, upon which this robotic top can skitter around.**

2. **A swarm of Robosapiens, a commercially available toy robot designed by Tilden that incorporates many BEAM principles.**

Robosapien, for example, contains seven motors, but its gearboxes are designed to allow powered movement along one joint to transfer over to neighboring joints without engaging the additional motor. So the Robosapien's torso will passively rotate back and forth in response to its powered waving of an arm.

Thanks to such optimizations, BEAM bots will likely be among the first to achieve power autonomy with the arrival of cheaper, more efficient solar cells and power storage technologies. This will fulfill one of Tilden's Laws of Robotics (laid down in response to Asimov's Laws): 1. A robot must protect its existence at all costs. 2. A robot must obtain and maintain access to its own power source. 3. A robot must continually search for better power sources. (Or, as they are more widely recited: 1. Protect thy ass. 2. Feed thy ass. 3. Get thy ass to better real estate.)

Basic BEAM Circuits

BEAM prides itself on minimalist design and an anyone-can-play openness. Even so, there are dozens of basic BEAM circuits, far too many to explain here. But to give you a taste of BEAM electronics, let's look at two fundamental circuits: the Solarengine and the bicore.

The Solarengine (SE): A *Solarengine* is a power circuit built around a solar cell and a storage capacitor. The solar cell trickle-charges the cap, and when the amount of stored energy is high enough, the cap dumps the charge into the circuit to power the robot.

Different types of Solarengines are defined by the method they use to trigger the dump. Type 1 SEs use a voltage-sensing trigger, Type 2 SEs use a timer, and Type 3 SEs monitor the current that flows into the cap and trigger the dump once the capacitor's charge-rate begins to slow down as a result of getting full. A fourth and final flavor of SE is the nocturnal variety, which reads ambient light levels and turns on the power once darkness descends on the robot.

A Type 1 Solarengine provides the basis for the Trimet and Solarroller project later in this issue *(page 76, with an annotated circuit diagram on page 79 that shows how its basic Solarengine works*). In this design, the voltage trigger is a 1381 chip, a three-pin component that opens the connection between the power and output pins when there's a 3.4 volt (or other voltage) difference between the power and ground pins. This sends the stored capacitor power into a motor.

The Bicore: Many of the most interesting BEAMbot behaviors are controlled by "nervous nets." These are simple control circuits that emulate low-level peripheral nerves in a spinal column. Each "neuron" within these circuits is typically composed of a timer sub-circuit (a resistor and capacitor) combined with one or more inverters (logical NOT gates). Because a 20-pin inverter chip can house eight gates, one inverter chip can simultaneously support several such neurons.

Photography by Jim Mullins (left) and Wow Wee Toys (right)

A one-motor walker I built from a 74HCT240 bicore and a servomotor with its control electronics removed. The bicore creates an oscillating signal that generates a back-and-forth walking gait.

Photograph by Jay Townsend/Primal Design

The most basic nervous net has two neurons, and is therefore called a *bicore*. When the two neurons feed back to one another, they can create an oscillating signal to generate a legged robot's walking gait. Bicores can also control motors for steering and LEDs for signaling, like in Dave Prochnow's Panzeroids project on page 58.

I made a simple walker using a Tilden-designed bicore built from a 74HCT240 eight-gate inverter chip. In making this two-neuron oscillator, you solder four gates of the inverter chip together. Hook this simple controller up to a gearmotor, and the back-and-forth signal makes the little critter walk and even climb over low objects.

To take it further, you can chain several bicores together in master-slave configuration, creating a multi-motor walker. The network's master bicore sets the rhythm and sends the signal down to the next bicore. Resistors chaining the series together set the time delay, determining the walker's gait.

You can also attach a touch sensor bumper to a bicore, and use it to invert the direction of power to the motors. This will make the walker back up when it runs into any objects. Combine a bump sensor bicore with a two-bicore, two-motor walker and you have a very cool little bugbot that can navigate a space and negotiate small obstacles. Not too bad for a couple of gearmotors and a handful of cheap analog parts.

Getting Started in BEAM

For more information, check out Solarbotics' *BEAM from the Ground Up* online tutorial, at solarbotics.net/bftgu. Also check out the rest of the Solarbotics site; it has massive amounts of how-to articles, circuits, component datasheets, and a BEAM encyclopedia and "bestiary." The commercial side of the site offers BEAM parts, books, kits, and other robot-related products. Any one of their inexpensive kits is a great way to get your feet wet.

Also see Dave Hrynkiw and Mark Tilden's indispensable book *Junkbots, Bugbots, and Bots on Wheels* (Osborne, 2002), and my own book, *The Absolute Beginner's Guide to Building Robots* (Que, 2004).

Gareth Branwyn writes about the intersection of technology and culture for *Wired* and other publications, and is a member of MAKE's Advisory Board. He is Cyborg-in-Chief of Streettech.com.

Panzeroids

Become a desktop general with these battlin' bot tanks.

By Dave Prochnow

WELCOME TO THE BATTLEFIELD OF the future, where robot gladiators fight to decide the political power of their human commanders. Robots have replaced human soldiers, and human deaths on the field of battle are a historical memory. Yeah, right — dream on!

Well, actually, this vision of tomorrow is here today, scaled down a bit. Behold: Panzeroids, robot tanks that battle to the death — an LED-flashing death. The field of conflict? Your kitchen floor. The scale may be small, but the combat is real. Best of all, Panzeroids won't cost you the national debt; you can easily build two dueling tanks for less than $50.

To build a Panzeroid, you start with a stock motorized model tank kit. Then you hack in a type of control circuit called a "bicore" and an infrared (IR) LED emitter flasher. What you get is a robotank that can independently locate, track, and "kill" another similarly equipped tank. Here's how I built mine.

A Robot Hacker's Delight

For tanks, I used two Leclerc French Army Main Battle Tank (MBT) kits from Academy Models. I bought them for just $16 each, and at that price, these kits represent one of the best bargains in robot hacking. You get two motors, complete with gearboxes, mounted in a neat, self-contained chassis. Even if you don't build a tracked robot, you can use the kit's motor/gearbox combination for numerous other projects.

The first step is to assemble the plastic body of your model tank, as described in the kit's instructions. But you should stop just before the part where you attach the remote control cable to the motor gearboxes. This is where the hacking begins and the Panzeroids are born. Instead of connecting to the tethered remote, each motor gearbox will connect to your robot's bicore "brain."

Photography by Dave Prochnow

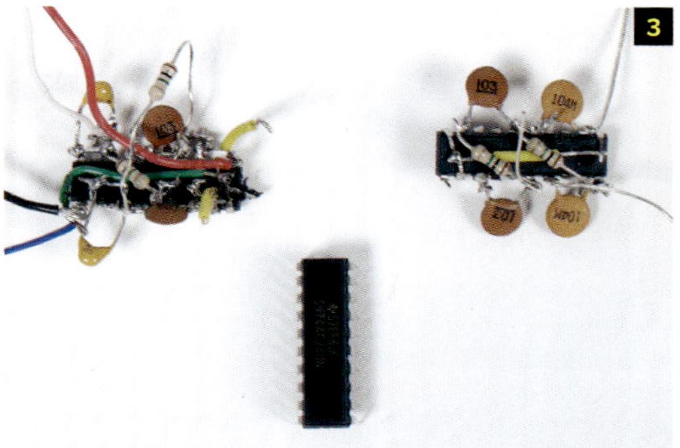

1. The Academy Models Leclerc MBT kits include motors and gearboxes, and represent a terrific bargain for robot hackers.

2. The Leclerc model tank fits together without cement, so it's easy to access the batteries inside.

3. Each Panzeroid tank has two suspended bicore "brains" based on a TI 74AC240 chip.

Two-Track Minds

The next step is to build your robots' brains. There are many different and equally valid ways for building bicores, and I won't explain how to do it here, but check the "Plans and Sources" list at the end of the sidebar on page 61 for where to find some good instructions. You'll need two suspended bicores for each tank. One will control the tank's movements by dividing and adjusting the power (0V to +5V) between each of its two motors, and the other will control the IR LED flasher "gun" that's mounted inside the tank turret.

The bicores I built are based on the 74AC240 Octal Buffer Inverter IC. In addition to these chips, you also need a few garden-variety capacitors, resistors, photodiodes, and LEDs. They're simple circuits to put together, and all told, the total cost for two tanks' worth of bicore brains was about $6.

Flasher vs. Motor Bicores

Note that for the Panzeroid flasher controller bicores, the suspended resistance is a simple resistor, as shown in most bicore schematics. But the motor control circuits have one substitution: instead of a resistor between the two cores, you use a pair of photodiodes connected in series and oriented in opposite directions. These twin photodiodes function like a resistor, and also alternately modulate the power running to each tank motor, depending on the level of IR they receive. This is what makes the tank locate and steer toward the strongest IR signal.

Before soldering together my bicore circuits, I tested them out on a breadboard, temporarily attaching two red LEDs to the control voltages. Each bicore has four connection points: power, ground, and the two control voltages. With the 74AC240 chip-based design I used, these correspond to pins 20, 10, 9, and 12, respectively.

Rebraining Procedure

Once you have your bicores built, it's time to hook them in. Each Academy Leclerc model kit comes with a tethered remote control. We'll use the remote's wires to connect the motors to the bicore brains. Open up the remote unit and desolder the four control wires (blue, yellow, red, and white) from the circuit board.

Take the four wires you just desoldered, and solder them to the four terminals on the tank's two motors. Connect the white wire to the top (ground) terminal of the tank's left motor and the blue wire to the top terminal of the tank's right motor. Solder the remaining red and yellow wires to the bottom (power) terminals of the tank's left and right motor terminals, respectively. Then cut off all but about 18 inches of the ribbon cable, and separate and strip the other ends for connecting later.

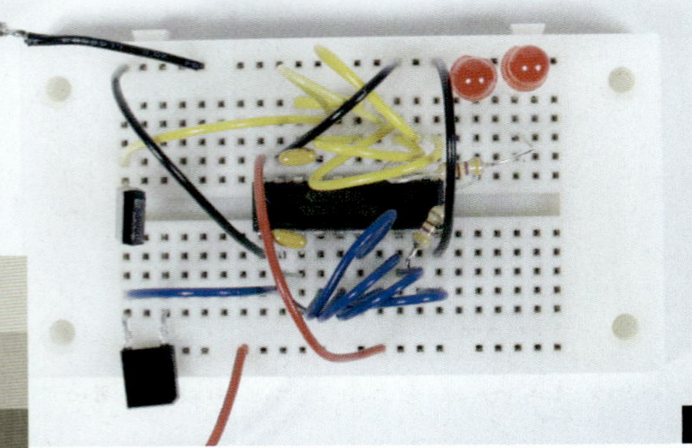

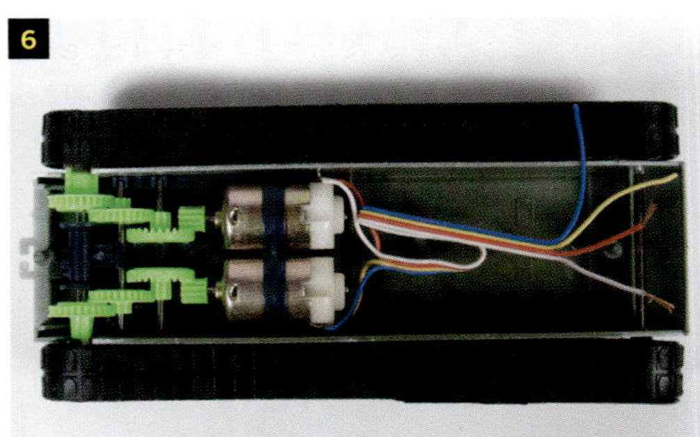

4. Basic bicore (from ***Junkbots, Bugbots, and Bots on Wheels***, **p. 288) assembled for testing on a breadboard. Resistor-LED pairs simulate the motors, and the black rectangles at left are photodiodes.**

5. **Inside the remote, four wires to desolder.**

6. **Wires from the remote soldered to the motors. The other ends connect to the motor bicore in the turret.**

7. **Master bicore brains inside tank turret (also from *Junkbots*, p. 278). One connects to two IR LEDs on the left side, and the other connects to the motors and two photodiodes on the right side.**

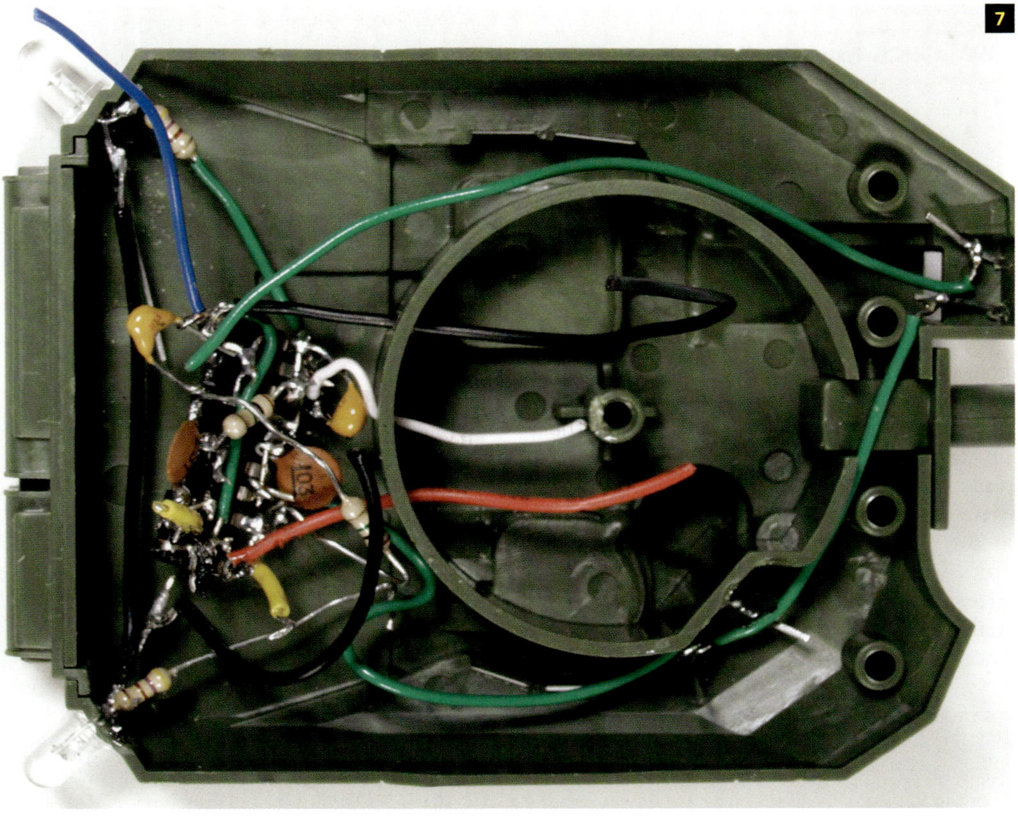

Try to squeeze your two bicore brains into the tank's turret. Connect the bicore with the resistor to the IR LEDs positioned along the left side of the turret, and put the twin photodiodes from the motor-control bicore on the opposite, right side. Then run some wires up the turret from the battery compartment in order to hook the two bicores up to power. Connect the power pin on each IC (pin 20) to the positive battery terminal, and the ground pin (pin 10) to the negative battery terminal.

Decide how you want to position the photodiodes and LED flashers. For optimal IR light sensitivity, the two photodiodes should be spaced away from each other. I installed them poking out of different port details in the turret, one next to the barrel and the other on the side. I put the LED flashers on the rear corners of the turret.

Hook up the bicores' outputs. With the IR bicore, wire its IC's output pins (pins 9 and 12) to the anodes of the two IR LED emitters. Then connect the LED cathodes to ground. For the motor control, with the twin photodiodes, solder the left motor power (red wire from the remote cable) to pin 9 and the right motor power (yellow wire) to pin 12. Connect the motor grounds (white and blue) to ground, via pin 10 or the negative battery terminal. Finally, join the turret to the tank's body, connect the battery, and let the tank run on a test drive. The motors should steer the tank around as it seeks out an IR light source.

Patton vs. Rommel

Now, let's watch two Panzeroids battle it out. Each will pulse IR from its flashers in back, while simultaneously trying to drive toward another IR light source — namely, the flashers on the other tank.

The tanks will steer around, reacting to variations in flash intensity, battery strength, ambient light, visual obstacles, and other factors. When one tank senses the other, it will steer toward it. The first tank that locks on long enough to drive up to and ram the other tank is declared the Panzeroids champ. Die, robot, die!

Note that strong sunlight can adversely affect the performance of your Panzeroids competition. Therefore, if you don't want your Panzeroids trying to fight the sun, you should confine your combatants to a dimly lit coliseum.

Dave Prochnow is a frequent contributor to MAKE and an award-winning author of 25 nonfiction books, including the bestselling *PSP Hacks, Mods, and Expansions* (McGraw-Hill, 2006). You can learn more about his projects at pco2go.com.

More on the Bicore

The bicore was invented by Mark Tilden, the father of BEAM Robotics and the Robosapien toy (*see page 54*). During early research into using electronics to simulate animal neural networks, Tilden developed the Nv neuron, a simple circuit that uses a resistor-capacitor (RC) timer and an inverter to slow electrical signals down, so that they propagate at meatware speeds. This lets electronic circuits time-slice changes in the real world at about the same rate that biological systems do, giving them a "reflex" speed that works well for controlling inertia-bound physical objects like motors and wheels.

Tilden and others combined these Nv neurons into networks, which he originally termed Very Slow Propagation Artificial Neural Systems (VSPANS). Now they're called "nervous networks," and people refer to them by the number of Nv "cores" they contain. A network with two Nv neurons is a bicore, one with three neurons is a tricore, and so on.

Our Panzeroid control circuits have two cores built using one chip, connected in series, with a resistor suspended between them. This is a "suspended bicore" design, and the resistance that separates the cores provides a "virtual ground" that swaps circuit action back and forth between the two. This creates an oscillator that can drive many types of circuits, including the Panzeroids' motors and IR flashers.

Plans and Sources

You can learn how to build bicore circuits from either of the following sources:

* **BEAM Online: makezine.com/go/beamonline**

* ***Junkbots, Bugbots, and Bots on Wheels* by Dave Hrynkiw and Mark W. Tilden (McGraw-Hill, 2002). This is the must-read bible for anyone interested in BEAM robotics.**

Or, you can buy a Bicore Experimenters PCB from Solarbotics (solarbotics.com, product #BEP, $35).

Model Tank Sources

If you have a hankering for hacking motorized models into viable robots, here are three manufacturers who produce acceptable tank kits. You can order kits online from Squadron Hobby (squadron.com) and Tower Hobbies (towerhobbies.com).

* **Academy (academyhobby.com): 1:48 scale Korean kits; best for general robot hacking.**

* **Micro-X-Tech (dragonmodelsusa.com): 1:72 scale Hong Kong kits; tiny, detailed kits that require advanced hacking skills.**

* **Tamiya (tamiyausa.com): 1:35 and 1:16 scale Japanese kits; outstanding quality for high-performance hacking.**

Bots in a Snap

The nuts and bolts of Lego robot design. By Michael Rosenblatt

WHEN I BUILD LEGO ROBOTS, I GET out the drill. I have been doing this since 2002 when I had the job of building a robotic search and rescue demonstration for the U.S. Office of Naval Research. This job was commissioned to develop an immersive activity exposing trainees to the very real challenges of robotic search and rescue. With an aggressive budget and schedule, I decided to use a combination of Lego and custom-fabricated materials to build the working search robots. The two robots had to be durable and powerful enough to climb steep inclines and simulated staircases, have working pan-tilt cameras, and, particularly challenging, carry a smaller "scout" robot onboard.

The Lego purists will cringe, but I adopted a liberal approach to construction, using any materials and fabrication methods available to achieve functionality and scale. This article shares ten tips I learned during this project that I think you will find useful when applied to your own contraptions.

1 Use vertical through-bolts for stronger structures.

Why: You can reinforce structures by adding more Legos, but bolts are stronger and use less space.

Materials: #4-40 size machine screws, nuts, and washers.

Procedure: Drill through Lego Technic studs with a ⅛" bit. With a washer, slip a 4-40 screw through the holes. Snug, but don't over-tighten, as you might crack the Lego plastic.

2 Use machine screws to mount non-Lego materials on to your creations.

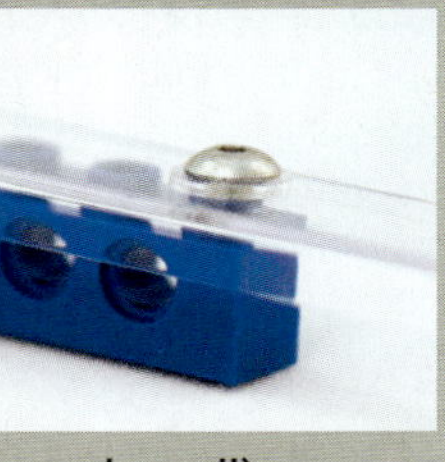

Why: Add a platform as large as you want for your Lego bricks, as well as reinforcing the structure.

Materials: #6 machine screws, #6 washers, and any flat sheet material (⅛" acrylic works well).

Procedure: Drill holes in your sheet material at the appropriate places for your design. Drill through the mating Lego studs with a ⅛" bit. The larger machine screws will securely thread into the hollowed Lego studs to mount your sheet. Be careful not to strip the holes by over-tightening.

Photography and illustrations by Robin Tafel

3 Use art-store aluminum tubes to make linear slide mechanisms.

Why: Make longer Lego slide mechanisms than you can with the longest plastic axles.
Materials: Metal tube or rod 3⁄16" diameter, available at many art/craft stores and hobby stores.
Procedure: Cut the tube/rod to desired length. Actuate with Lego flat-gears, a string/pulley mechanism, or springs.

4 Install cheap rotary sensors (encoders, potentiometers, rotary switches).

Why: Use a multitude of inexpensive sensors that interface directly with your Lego projects.
Materials: Any 1⁄4"-base, 1⁄8"-shaft rotary sensor, and 16T Lego gear.
Procedure: Expand the hole in a Lego Technic beam with a 15⁄64" bit. Force-thread the sensor into the enlarged hole, being careful to keep the shaft perfectly centered in the hole. Press the 16T gear onto the shaft. If the gear slips on the shaft, use a little plastic epoxy.

5 Make your own Lego encoder with an LED and photosensor.

Why: It's a cheap, compact rotation sensor.
Materials: T-1 3⁄4 size LED, T-1 3⁄4 photo-diode or photo transistor, Lego pulley, and Technic beams.
Procedure: Press LED and sensor into the beams. Allow the pulley to spin. May require an analog microcontroller input to read, as the holes do not fully block the LED light.

6 Mount a servo with horizontal axis to drive Lego gear trains.

Why: Actuate Lego gear trains with your servo.
Materials: R/C hobby servo, #4-40 screws, nuts, washers, Lego pulley, Lego Technic beam, and two-sided foam tape.
Procedure: Tape the servo onto smooth-top Lego plates. Reinforce with vertically locking Lego structure. To drive Lego gear trains, make sure the servo's output shaft is aligned with the hole of a Lego Technic beam. Create a shaft coupling by screwing a pulley to the appropriate servo horn.

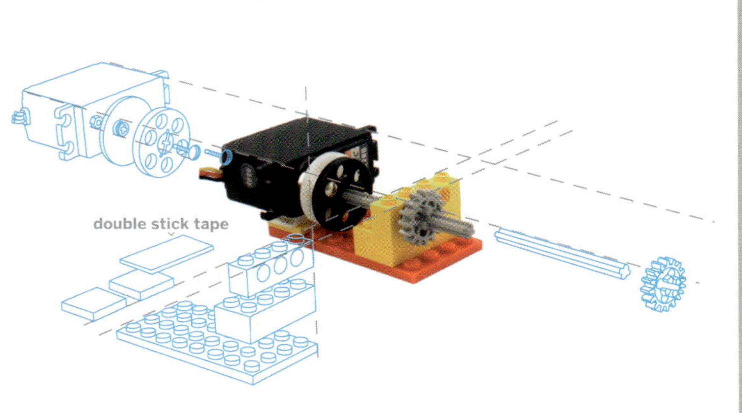

7 Mount a servo with a vertical axis, to drive mechanisms via linkages.

Why: Servo mount to actuate basically anything.
Mounting materials: #4-40 screws, nuts, washers, servo horns, #2-56 ball swivel or ball joints, and #2-56 threaded rod.
Mounting procedure: Construct a structure around your servo. Refer to tip #2 for screw mounting.
Push-rod procedure: Mount one ball swivel onto servo horn with screw. Mount the other ball swivel onto any Lego part you wish to actuate.

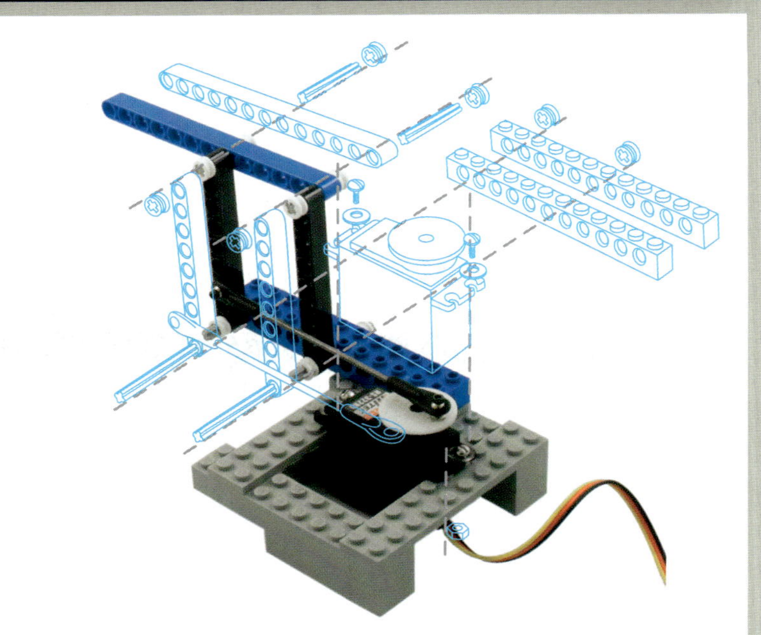

8 Mount PCBs and electronics to Lego models.

Why: Save space by removing unnecessary electronics enclosures.
Materials: Hot glue and dispenser, or 3M double-sided foam tape.
Procedure: When mounting PCBs on top of a Lego stud surface, use hot glue. When mounting on the smooth surfaces (sides and smooth plates), the 3M tape will work best.

10 Mount graphics on your Lego creations.

Why: Add graphics to the smooth sides of your Lego models.
Materials: Paper (photo paper or transparency), 3M spray mount, and inkjet printer.

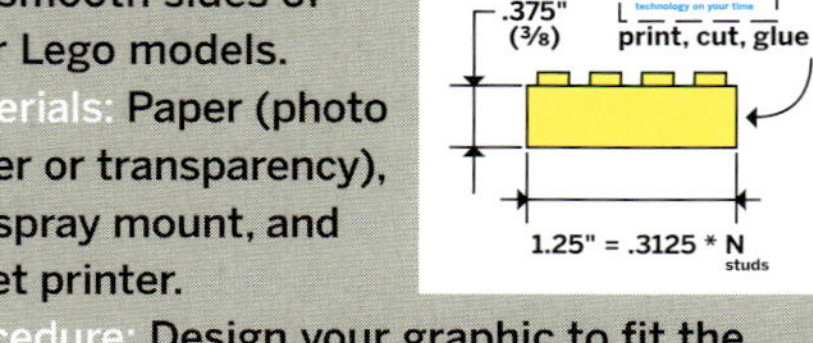

Procedure: Design your graphic to fit the dimensions of an individual Lego brick side or an entire surface. Print and cut out. Spray glue onto the back of the graphic and press onto your smooth surface.

9 Make acrylic or wood plates that fit with Lego studs.

Why: Make custom-sized plates to add functionality and structure to your Lego constructions.
Materials: 1/8" acrylic, aluminum, or basswood sheet.

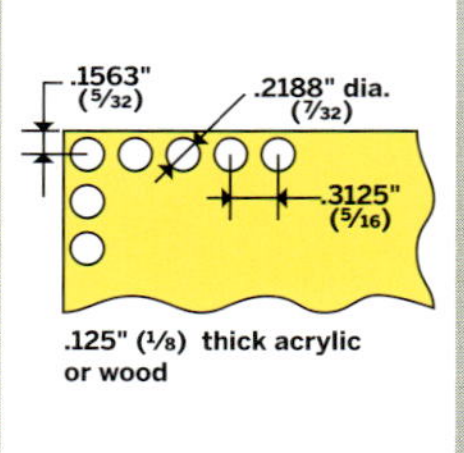

Procedure: Design your pattern in Adobe Illustrator or a 2D CAD package, based on the dimensions given in the drawing. Then fabricate your plate with a drill press, scroll saw, and/or band saw, or send the job to a fabrication shop that can make it with a laser or water-jet cutter.

For a list of resources, visit makezine.com/06/lego.

This article was inspired by *The Art of LEGO Design* by Fred G. Martin (1995): handyboard.com/techdocs/artoflego.pdf.

Michael Rosenblatt works as a program manager at Apple Computer and is a graduate of the MIT Media Lab. He is the founder of inventionDB.com, which assists designers and engineers in quickly documenting projects.

Cogs and Cocktails

Meet the drink-serving, drunk-driving droids at Roboexotica.

By Cory Doctorow

Photograph by Jacob Appelbaum/jacob@appelbaum.net

THE MOJITO-MIXING ROBOT HAS A long, steel spiral slide through which limes tumble, landing in a machine that squeezes their juice into a glass. Then, a clanking chain drive moves the glass through a series of stations: a mint dispenser, a muddler that pounds the leaves in with the lime juice, thence to rum, sugar syrup, ice, and soda dispensers. The whole thing is festooned with blobby welds, snaking cables, and exposed logic.

For three years now, Robert Martin, an ex-CNC technician, has brought his mojito robot to Vienna's annual Roboexotica convention (roboexotica.org), where he and other amateur roboticists show off their motley, semi-functional "cocktail robots" — droids that mix and pour drinks, light cigarettes, and above all, inspire conversation. Participants come from all over the world and every level of expertise, from Silicon Valley professors to a guy who simply exhibited a tank of beer with a submersible in the suds.

The mojito robot never successfully prepares a drink, but it always draws a crowd. It's so pretty, so impractical, and so teasing in its potential that watching it not work is more fun than watching it work would be. "It's about beautiful failure and interactivity," says Johannes Grenzfurthner of Monochrom, a Viennese tech/arts group that co-organizes Roboexotica along with another tech-creative outfit called Shifz.

Magnus Wurzner, the ringleader of Shifz, leads me past two 7-year-olds remotely piloting a robot to bump into people and offer them peanuts. Down the hall, a roaring slot-car race is controlled by EEG contacts — the drunker the drivers get, the faster their cars go. At a chrome bar, an animatronic drunkard vomits copiously into a cocktail glass.

Behind the bar looms Thunder One, a giant commercial drink-mixing machine with an LCD interface and a snake's nest of tubes that require regular flushing to clean them. Wurzner points and shouts over the din: "Thunder One is what cocktail culture will come to if we don't intervene. It's all about efficiency and not about aesthetics or irony. That's what cocktail robots and cocktail culture are for."

Cory Doctorow (craphound.com) wrote "A House Divided" on page 26.

The Vex Robotics Design System

Versatile, powerful design raises the bot in prefab robotics construction kits. By Gareth Branwyn

GADGET GEEKS HAVE COME TO EXPECT that anything Segway inventor Dean Kamen is associated with is going to be innovative, possibly a bit over-hyped, and undoubtedly expensive. Such is the case with the Vex Robotics Design System (vexrobotics.com). Offered exclusively at RadioShack, Vex grew out of Kamen's FIRST Robotics Competition (www.usfirst.org).

Where the Lego Mindstorms Robotics Invention System grew, in part, out of research and robotics competitions at MIT, Vex has origins in Carnegie Mellon University's Robotics Institute. The result is, like Lego, a friendly and open-ended user experience with a wealth of online resources.

Even at $299, the Vex Robotics Design System Starter Kit is reasonably priced for what you get (the add-ons, however, are another story). Its handsome carrying case comes filled with over 500 parts, including three gear motors, one servo, a variety of tires and hubs, 14 different gears, a microcontroller module, an RF receiver module with a replaceable frequency crystal, and a surprisingly decent 6-channel radio transmitter.

If Mindstorms is the robotic answer to Lego, then Vex is the Erector/Meccano analog. The main building components are stamped steel beams, rails, angle pieces, and panels, with attendant nuts, bolts, washers, and other hardware. The structural pieces are Swiss-cheesed with square holes to provide a maximum number of attachment points. The Starter Kit also includes four sensors: two limit switches and two bump sensors.

Your first Vex project: assemble the binder. The *Vex Inventor's Guide* that comes with the kit is a thick D-ring binder and a bundle of loose-leaf pages that you have to organize into it, including instructions for building a starter bot and other kit details. But each tabbed section has an excellent beginner's tutorial on mechanical engineering, electronics, or other concepts used in robot building. The guide is thoughtfully modular, and as you buy add-on subsystems, they come with hole-punched documentation that you add to the proper tabbed section of the binder. Guide sections are all color-coded with the kit parts themselves — silver=structural materials, green=motion, red=sensors, and so on — and this color-matching helps newbie builders understand the mechanical and electronic subsystems and how they interrelate. Product boxes for the add-on subsystems and components are similarly color-coded.

If Mindstorms is the robotic answer to Lego, then Vex is the Erector/Meccano analog.

The Add-Ons Add Up

It is in the add-on systems that the real cost of the Vex system reveals itself. While the Starter Kit does give you a lot of great components for the money, you don't get a full, true robot experience until you add a number of parts. Beyond using the included remote control radio to operate your creations (not very bot-ish), the Starter Kit's microcontroller can be programmed only minimally out of the box, using banks of jumper blocks and little keys that turn on pre-wired routines. To do true programming, you need to purchase the Programming Accessory Kit ($99). Also, the Starter Kit ships

Photography courtesy of Innovation First, Inc.

Here's one sophisticated design, the Vex Tomahawk from VexLabs, which originally designed the Vex system. It's a four-wheeled vehicle that uses two servos and differential power to lean into turns, up to 30 degrees, with Segway-like ease.

VexLabs' remote controlled (no microcontroller) Ping-Pong-Bot uses an accumulator arm to scoop up and carry up to seven ping-pong balls, which it can then fling into the air one by one with a spring-loaded firing mechanism.

with a battery holder, but you have to supply 14 AA NiCad rechargeables yourself, or else buy the Vex Power Pack ($50). If you want to use your bot and RC transmitter in competition against another Vex robot, you need another crystal set with a different frequency from your competitor ($50). And limit switches and bump sensors get old pretty quickly. To add other sensors, you'll need to buy them, at a cost of between $20 and $40 per sensor kit. So a complete user-programmable, competition-ready Vex system can approach $500 or more.

While the true cost of entry may be higher than RadioShack's promotional Starter Kit pricing might imply, things do start to get interesting once you add the Programming Kit and additional sensor sets. When the Lego Mindstorms line was introduced, Lego was surprised to see it embraced not only by kids as an educational toy, but also by robot hobbyists, serious robot developers, and prototypers. But these groups discovered some drawbacks in Lego-based robotics, including the relative fragility of the interlocking bricks, and the limitations of exclusively brick-based design. Vex components offer many more structural configuration options than Lego. Everything is fastened with conventional nut-washer-bolt hardware, and you can bend and cut components. This allows the Vex system to scale from kid's building set all the way up to professional development tool.

The Versatility of Vex

To demonstrate the Vex system's versatility, Grant Imahara, of *BattleBots* and *MythBusters* fame, built a working replica of iRobot's military PackBot using Vex, including the Starter Kit, Programming Kit, Tank Tread Kit ($30), and the Ultrasonic Range Finder Kit (also $30). An article describing the build can be found at makezine.com/go/vexrobo.

Imahara was impressed at how quickly he could construct the Vex PackBot, and the degree of sophistication it could achieve. The dual-tank-tread robot used a remote, semi-autonomous co-control system which allowed it to make some of its own decisions about how to negotiate obstacles.

Jamie Hyneman, also known for *BattleBots* and *MythBusters*, repurposed the Vex system's radio control system and microcontroller unit to create a series of full-sized, computer- and remote-controlled bumper cars for a corporate client — ahead of time and well under budget.

The Vex system offers a great "out of box" experience, and more robust components than Lego. With its versatile microcontroller and easy graphical programming environment (based on the EasyC language), Vex combines construction-toy ease with grown-up computer power. Add the already large user community, thanks to FIRST, and there's monstrous potential for this system. Being able to snag parts at any RadioShack doesn't hurt either.

Hacking the Vex Microcontroller?

As with its Mindstorms forebears, it didn't take long before builders were cracking open the Vex system's microcontroller unit (MCU) and trying to figure out how to make a programming cable that could talk to it. The Vex is basically a simpler, cheaper, commercial version of previous FIRST robotics building components sold by Innovation First (ifirobotics.com), and the Vex MCU uses the same Microchip PIC 18F8520 microprocessor found in the older FIRST Robotic Control (FRC) systems.

This means that if you can build a serial interface cable for the Vex MCU, you should be able to load code onto it using existing FRC software tools such as the IFI Loader (free) and Microchip's MPLAB IDE and C18 Compiler (free 60-day demo).

As of this writing, no one has done this yet, and discussions of the possibility are frowned upon on official Vex and FIRST sites. But it's only a matter of time before such programming options become available. There is a preliminary discussion on investigating the MCU's serial pin-outs at makezine.com/go/vexhack.

Gareth Branwyn wrote "A Beginner's Guide to BEAM" on page 54 and the "Two BEAMbots: Trimet and Solarroller" project on page 76.

Mini Mars Rover

A wireless remote control camera on wheels. By Tom Zimmerman

Photography by Tom Zimmerman

WHEN I GO TO SCHOOLS, I TELL THE students about the Mars Rover, a remote control robot with cameras that roams around Mars. Then I tell them they're going to drive one. Well, not exactly. I bring out a remote control truck with a wireless video camera on top and magnets on the front bumper. Then I spread tin cans on the floor and have students remote-control the truck to collect them – just like the Mars Rover, except with cans instead of rocks. Sounds easy, right? Wrong. They can only see what the camera sees, on a TV screen, and they soon find out that driving is a lot harder when their field of vision is as narrow as a video camera's.

You, too, can build the Mars Rover. Here's how to do it in a few hours for under $150. The main components are a remote control (R/C) truck and an X10 wireless camera and receiver. The only tools you need are a hand drill, diagonal cutters, a saw, and scissors. And surprisingly, the hardest part of the project is arranging the magnets.

TEST THE CAMERA

Before you start drilling, make sure the wireless camera works. Load 4 AA batteries into the camera's battery pack, and plug the wireless receiver into a TV using the RCV audio/video cable. (If the TV lacks separate video and audio jacks, plug its RF output into the antenna input.) Use the small slide switches on both the camera and wireless receiver to set each device to channel 1, and try other channels if you don't get a clear signal. If you are near a wireless

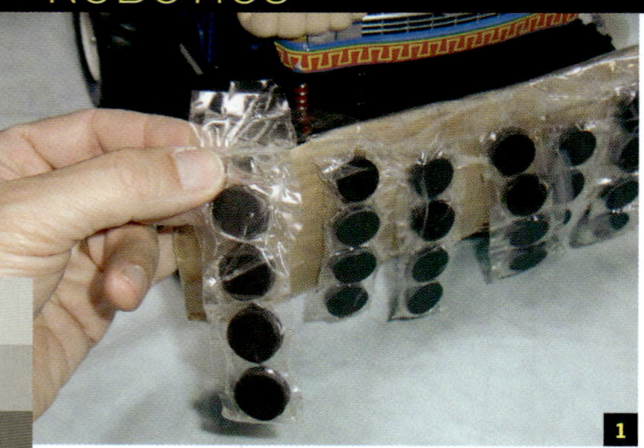
1

2

SUPPLIES

Wireless camera and receiver **X10 #XCAM2, $80**
Battery pack **X10 #XCam2 Battery Pack, $20**
RCA audio/video cable **12' RadioShack #15-1512, $15**
Remote control truck **Big enough to carry camera and battery pack (try RadioShack or a hobby or toy shop)**
Magnets (28) **RadioShack #64-1888, 5 for $2.60**
Diagonal cutters **RadioShack 64-2951, $5**
Silicone rubber sealer **RadioShack #64-2314, $3.20**
AA batteries (4) **For battery pack**
Hand drill and ¼" bit **For drilling into plastic and wood**
Cable ties (6 or more) **Less than ¼" wide and 6" long**
Clear 2" wide tape **At least 5'**
Cardboard **About 4"x10", enough to cover the truck's front bumper**
Scrap wood **At least 3"x6" to mount the camera to the truck**

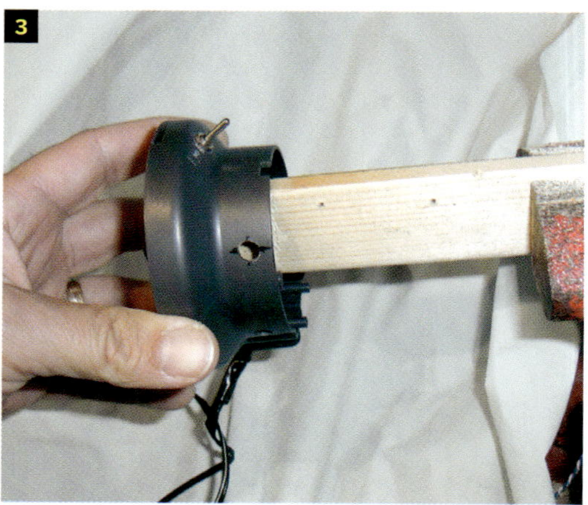
3

network (WLAN), which also operates at 2.4GHz, you may see white horizontal lines in the image. The 2.4GHz band is unlicensed, and devices operating on this band must accept interference. If this is a problem, move to another room or somewhere else.

Another thing to consider is communication range. The camera manufacturer claims a range of up to 100 feet, but as with all wireless communication, this depends on your environment. Outside gives the best range, and buildings with metal in the walls will give you the shortest. White snow in the video image, which is how analog signals like this degrade, means that you're reaching maximum range and should move closer. Once you lose the video, you'll be driving blind.

The camera's battery pack has a long cable, which you can either fold neatly and tie out of the way or shorten down. I cut mine down because I wanted to get to the wires anyway, in order to test the camera's voltage and power consumption, so I shortened the cable to about 2 feet. (Power consumption turned out to be about 1 watt, 16.5V x 65mA.)

CHOOSE THE TRUCK

When you shop around for the R/C truck, think about where you will attach the camera and battery pack. The truck I bought has railing in back where you can attach cable ties, and the cab's roof is hollow, so you can drill mounting holes into the plastic.

ADD MAGNETS TO THE BUMPER

Cut a piece of cardboard to fit the front of the truck bumper, about 4"x10", then punch pairs of holes in it and cable-tie it on. If your truck doesn't have a bumper, drill some ¼" holes in the front plastic, but make sure you don't drill into the turning mechanism or any other critical components.

The cardboard supports flexible, hanging strips of magnets that wrap against the cans to ensure a strong hold (Fig. 1). To make each strip, cut a 6" piece of tape and place 4 magnets lengthwise, starting at the bottom left side of the tape. This leaves about 2" of tape at the top, where you connect it to the cardboard. Carefully fold the right side of the tape over the left side, enclosing the magnets in the tape. I used 7 strips of magnets for my truck.

Finally, tape the top of each strip to the cardboard bumper. As the truck moves, the tape will allow the magnets to grab and hold a can. If cans slip under the bumper, move it lower to the floor.

MOUNT THE CAMERA

Before you commit to a mounting location, try the camera out at different places on the truck's roof and see what kind of image you get onscreen. Ideally, the view will be straight ahead, with the top of the bumper visible at the bottom of the image.

Mount the camera with cable ties, which are cheaper and easier than nuts and bolts, and don't require exact alignment. If the cable ties are too wide to pass through the camera's mounting holes, widen them with a ¼" drill.

After driving my truck around, I realized that I'd get a better view if I placed the camera higher and farther back. So I cut a piece of scrap wood to fit between the truck cab and the plastic railing, sanded the top corners, and then drilled holes to cable-tie the wood to the railing and the camera. I secured the block of wood to the truck bed with a bead of silicone glue. It took me a couple of tries to get the camera properly aligned — and if you look carefully, you can see a few extra holes — but the result was a better view of the floor and bumper (Fig. 2).

MOUNT THE BATTERY CASE

It's a good idea to mount the battery pack case upside down so you can replace batteries without removing the case from the truck. I put my case to one side of the truck's back railing. Make sure the power switches for both truck and camera are easily accessible. Label the camera's power switch, since it will be facing downward and there is no power indicator light to show that it's on.

Once you know where to place the case, mark and drill four ¼" holes into the plastic on the rim. Start with ⅛" pilot holes, since big drill bits tend to grab in plastic, and use a strip of wood to support the fragile lip while the case is being drilled (Fig. 3). Cable-tie the case onto the truck, and trim the excess length. If you make a mistake or change your mind, just cut the ties and try again.

ROLL IT!

Now your Mars Rover is ready to roll. Remember: The magnets will attract only tin cans, which contain iron, and not aluminum cans. The audio track will be noisy, picking up the sound of the grinding gears and the giggles of children as you drive by their feet.

Going Further

Here's a game for your Rovers and a few people: Place four cans in the middle of the floor and challenge teams to pick up as many as possible in one minute. One team member watches the screen and drives, while the other is "home base" and removes cans. The driver collects a can, drives it home, and returns for more, only viewing the screen — no peeking allowed!

For the advanced tinkerer, here are some more ideas to extend your fun:

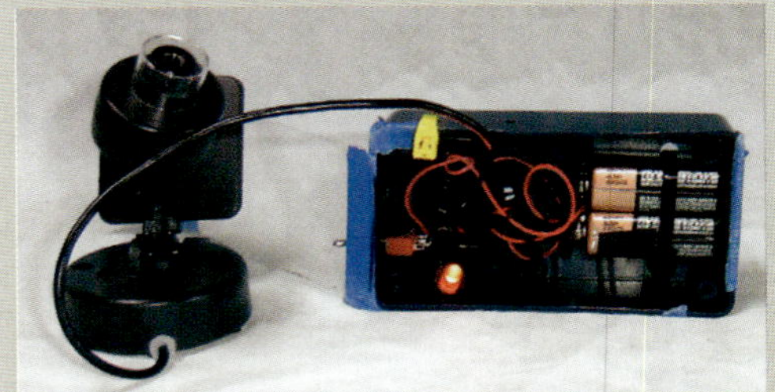

MAKE YOUR OWN BATTERY PACK **The camera says 12V, but I measured the battery pack at 16.5V, 65mA. At that current, a red LED drops nearly 2V. So I connected two 9V batteries and added a red LED in series with the camera, that acts as a power indicator.**

ADD A CAMERA **Since the wireless camera and receiver support four channels, you can put extra cameras on the truck, to look backward or to see cans on the bumper. Set the wireless cameras on different channels and use two receivers and TVs, or else switch between the channels/cameras on the receiver.**

ADD LIGHTS **The camera doesn't do well in the dark, but you can add some white LEDs or bulbs on the truck for illumination. For power, tap into the truck's battery. Even cooler, open the camera and remove the infrared-blocking filter (a little circular magenta piece of glass) to boost IR sensitivity, then use infrared LEDs as headlights, and see in complete darkness.**

ADD ULTRASONICS **This is way advanced, but you could build an ultrasonic range finder that converts distance to audible pitch, and send it through the camera's microphone. This would allow you to drive in complete darkness, like a bat, if a bat could drive.**

FLOATING CAMERAS **Instead of a truck, attach the wireless camera to a 3' weather balloon filled with helium. Use 100 feet of strong fishing line or string. I attached my camera to a 4"x12" piece of plastic, fashioned like a windmill tail, to stabilize it in the wind.**

Resources

X10 (wireless camera, receiver, battery pack): x10.com

RadioShack: See radioshack.com **for a store near you.**

Edmund Scientific: Two 3' diameter weather balloons, $26, item #3041755 scientificsonline.com

Hot Air

Build a do-everything manifold to control, dry, route, and use compressed air.

Gather the Necessary Parts

A brand-new compressor might impress your mother-in-law, but when you're ready to hit the big time, build this manifold system to get clean, dry air where you need it.

List of Materials

Most of the material you need for the Quick and Dirty Air Compressor Manifold can be purchased at a well-stocked hardware store.

[A] Compressor outlet Even small "pancake" compressors will operate a nail gun or small air tools, but don't expect to spray paint or repair a big rig. The size of the air outlet will vary by compressor size and output, so adapt it to ⅜".

[B] Flexible hook-up hose Compressors tend to jiggle around, so use a flexible hose to connect to your hard line manifold. Don't use a regular air hose! Use a special compressor connection hose that is rated at 200psi working pressure/ 800psi burst pressure.

[C] Ball valve In addition to turning off the air from the compressor, an inline ⅜" valve can be used as a manifold drain valve. A simple manifold won't accumulate much moisture, but when running longer lines, it's a good idea to have a drain at the lowest point.

[D] Galvanized steel lines Why use ⅜" galvanized steel pipe for air lines? PVC is a poor choice as it can shatter under pressure. Copper is ideal if you are not daunted by higher costs, increased labor, and great difficulty to make changes. I would use copper exclusively if I were a quadzillionaire with a plumber on staff. The pipe should have a slight decline down toward the drain valve to make sure moisture doesn't get trapped in the line.

[E] Regulator/moisture trap/filter Moisture is an unavoidable and damaging byproduct of compressing air. An all-in-one trap/regulator/ filter is cheap insurance to protect air tools from wet, dirty air pumped out at potentially damaging pressure.

[F] T-connector Use a ⅜" galvanized steel T wherever you want a drop.

[G] ⅜" male to ¼" female connector Maybe you will get lucky and find a ⅜" T connector with a ¼" outlet, but I had to adapt ⅜" to ¼".

[H] ¼" brass nipple Rather than a galvanized nipple, use a brass nipple with a hex in the center so you can use a wrench to get a tighter fit.

[I] 3-way air connector What kind of a savage doesn't love a brass 3-way air connector? Tidy and professional, it is an absolute delight.

[J] Combo coupler The bad news is there are three different "standard" compressor plugs — Industrial Interchange, Automotive Standard, and ARO Speed. The good news is Amflo makes a combo coupler that works with all three.

[K] ⅜" elbow Use to get around corners.

[L] Drop hose The ubiquitous yellow drop hose is found in every self-respecting auto shop in the world. I wouldn't dream of rebuilding a carburetor without one.

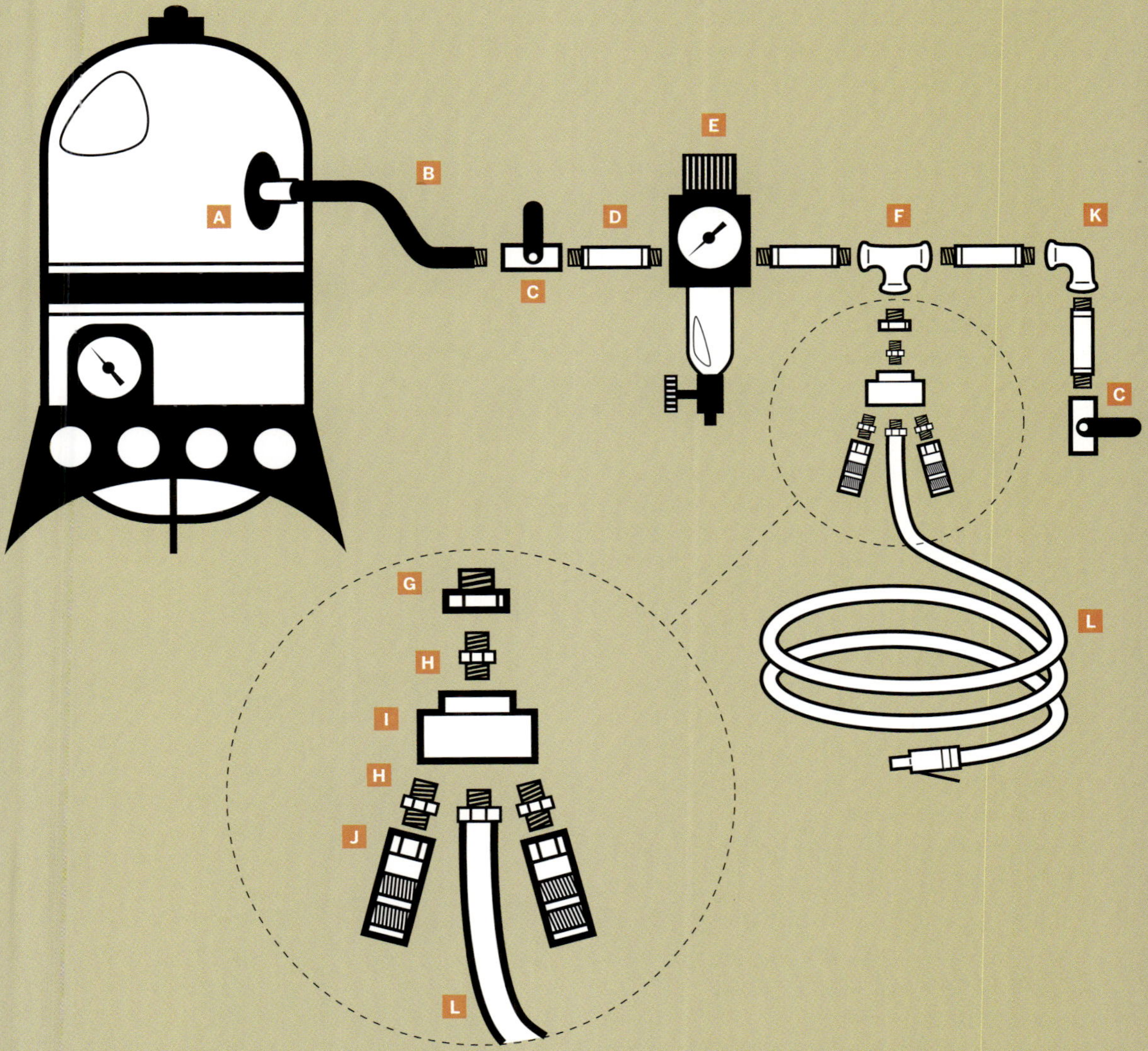

Illustration by Mister Jalopy and Mark Frauenfelder

Great Tire Gauge

Sure, there are digital tire gauges that speak, but I have enough friends and don't want to worry if I am spending enough time with my tire gauge. For $20, you can get a polished brass EZ-Air (accu-gage.com) with a standard feature list more impressive than the window sticker on a new Cadillac. Fill tires directly through the gauge and eliminate swapping between the fill hose and gauge — just watch the pressure rise as you fill!

Why not ½" or larger air lines? This ⅜" system is adequate for most tasks, but some jobs — like spraying auto paint — will require a more robust system with larger lines and specialized filters.

WARNING: **Compressed air dangers are no joke. Sounds far-fetched, but I know a fellow who is disabled due to a compressed air accident. Wear ANSI-approved safety glasses, read the manufacturer's instructions, and NEVER point a compressed air stream at friends, enemies, dogs, or even parakeets. Be safe!**

Q&D

How to Make an Airtight Threaded Connection

There are only two ways to put on Teflon tape — and one of them is wrong.

1 Holding the pipe with the open end facing you, start tape at 2 o'clock. Always wrap the tape clockwise around the threads. Otherwise, when you connect the female fitting, it will "roll up" the leading edge of the tape, resulting in an inferior seal.

What if the connections are reverse thread? OK, smarty-pants, reverse threads (a.k.a. left-hand threads) are taped counter-clockwise.

2 Why use pipe joint compound if you are already using Teflon tape? Some would say it's overkill, but the compound is a lubricant that allows you to turn the pipe an extra half turn tighter.

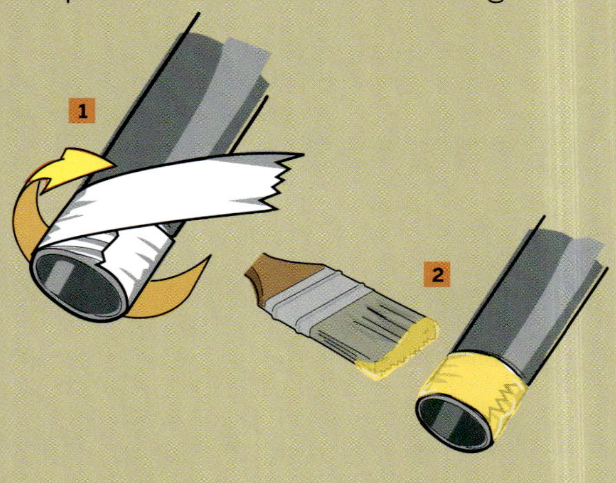

Now Make Yourself Useful

A whole world of nail guns, impact wrenches, cut-off wheels, pneumatic ratchets, die grinders, air hammers, and high-speed sanders will soon fill your dreams and make you forget about plugging power tools into electrical outlets. Behold the awesome power of compressed air!

Powerpuff Boy

If you don't require a heavy-duty air compressor, Hitachi's EC79 6-gallon pancake air compressor might fit the bill. The oil-free compressor, which weighs 60 lbs. and delivers 2.7cfm at 90psi, is perfect for finishing nail, staple, and other short burst work.
hitachipowertools.com

Illustrations by Tim Lillis

Mister Jalopy breaks the unbroken, repairs the irreparable, and explores the mechanical world at Hooptyrides.com.

Make: Projects

Bask in the glow of your soldering iron and build a geeky-cool bot battalion that follows the sun. If it's gloomy outside, just start practicing for RoboCup 2050 with a soccer-playing robot. Or reach new heights by delving into tensional integrity with tabletop sculptures that soar.

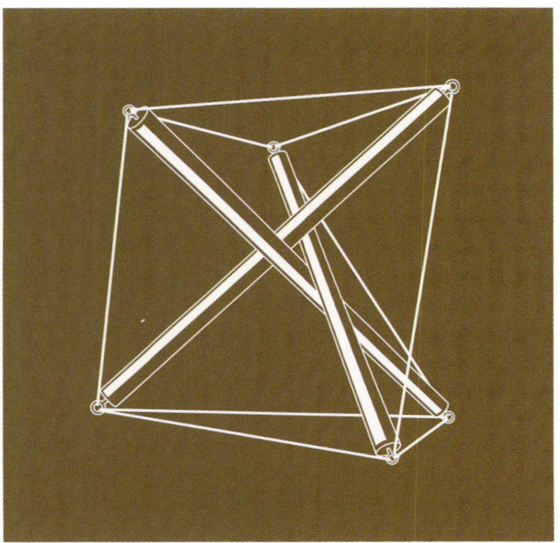

Photography by Douglas Ades

TWO BEAMBOTS: TRIMET AND SOLARROLLER

By Gareth Branwyn

Solder together one simple circuit and use it to control two very different solar-powered robo-critters: a little satellite that scoots and bumps around, and a mini cart that just keeps a-rolling until the sun goes down.

Set up: p.80 Make it: p.81 Use it: p.87

GO SOLARENGINE!

The low-tech, analog, dumpster-diving, and hack-friendly world of BEAM robotics (*see page 54*) has produced a bestiary of bot types, including Symets, Rollers, Walkers, Jumpers, Climbers, Swimmers, Flyers, and Crawlers. Many of these creatures can be powered and controlled by a Solarengine, a simple and popular BEAM circuit that draws energy from a solar cell and temporarily stores and dispenses it using one or more capacitors.

We'll make a couple of voltage-triggered Solarengine circuits, and then build them into two little bots: a Trimet, which looks like a satellite in orbit as it's moved around by a spinning, top-like base, and a Solarroller, which drives straight ahead in fits and starts. These light-sensitive critters will look cool and très geeky on your desk, as long as you can keep them from wandering off the edge (they're both active diurnally, and they don't have an off switch).

Gareth Branwyn wrote "A Beginner's Guide to BEAM" on page 54.

FAST REACTOR DESIGN

BEAMbots use simple circuits that interact with the world directly. Unlike control-freak robots, their brainless, reflexive reactivity is the whole point.

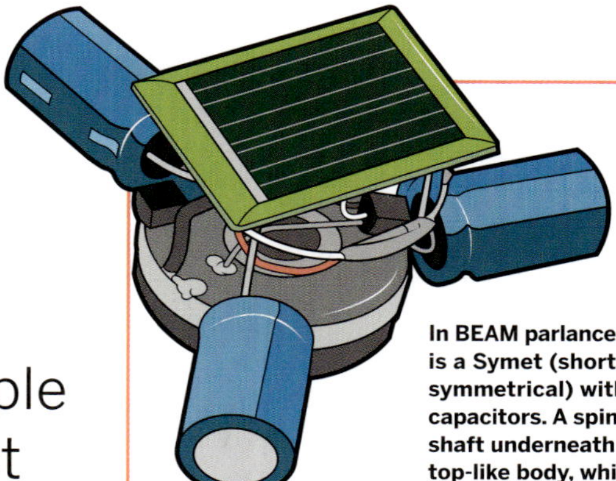

In BEAM parlance, a Trimet is a Symet (short for symmetrical) with three capacitors. A spinning drive shaft underneath pulls its top-like body, which bumps around any obstacles.

Solarrollers are little solar-powered race cars. At BEAM and other robot competitions, builders pit Solarrollers against each other in a kind of robo-mechanical Pinewood Derby.

Our Trimet and Solarroller bots are based on a voltage-triggered (Type 1) Solarengine. These circuits collect energy from a small solar cell, and periodically release it when there's enough stored up to actually do something, like run a motor.

The solar cell converts light into electrical energy, slowly juicing up the capacitor (or multiple capacitors).

The capacitor collects and stores a voltage, which discharges whenever the circuit is completed between its two terminals.

The 1381 voltage trigger measures the voltage across the capacitor, and sends a trigger signal once it's high enough (2.4 volts with a 1381-G trigger).

When the base pin of the 3904 transistor receives the trigger signal, it completes a connection that allows the capacitor's power to discharge through the motor.

The motor runs intermittently, whenever it receives a power dump from the capacitor.

During discharge, current flows to the base of the 3906 transistor. This takes the 1381 trigger offline, allowing it to reset, and routes current to the 3904 base, which keeps the motor circuit flowing until the cap is fully discharged.

The 2.2kΩ resistor reduces the voltage to the 3906 base pin, so it diverts less power away from the motor during discharge. This makes the circuit more efficient.

+ Solarengine schematic: makezine.com/06/beambots

Illustration by Timmy Kucynda

SET UP.

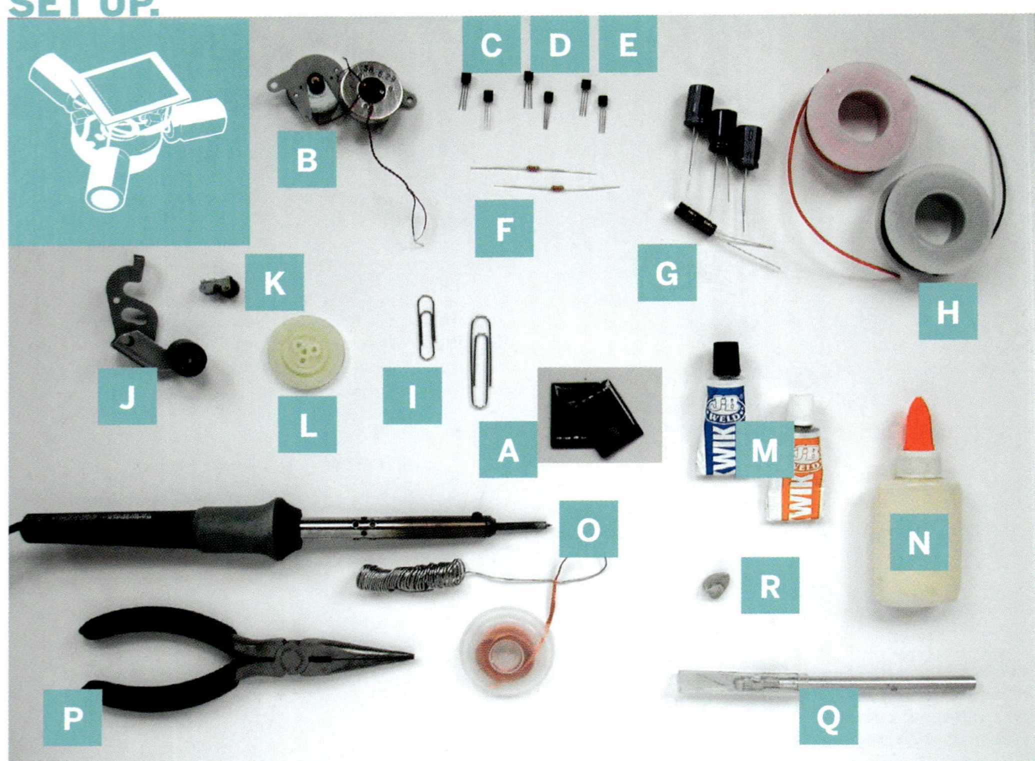

MATERIALS

The following parts will build two Solarengines. Just get one of each if you're only building the Trimet or the Solarroller, but not both. All part numbers refer to Solarbotics (solarbotics.com):

[A] 37x33mm polycrystalline solar cell, part #SCC3733 (2)

[B] Cassette motor (2) From an old Walkman or other player, part #MCM2

[C] 1381-G voltage trigger IC, part #1381-G (2)

[D] 2N3904 ("3904") NPN transistor, part #TR3904 (2)

[E] 2N3906 ("3906") PNP transistor, part #TR3906 (2)

[F] 2.2kΩ resistor (2)

[G] 4700µF capacitors (4) Or use three 4700µF capacitors (for the Trimet), and 1 "supercap" such as a 0.33F Gold Capacitor, part #CP.33F (for a higher-performance Solarroller)

[H] Hook-up wire, red and black 24-gauge stranded

[I] Paper clips (2) one small, one large

PARTS YOU'LL ONLY NEED FOR THE SOLARROLLER:

[J] Pinch roller and arm from a VCR, or similar Smooth rubber roller, about 5/8" in diameter and 5/8" wide

[K] Pinch roller and arm from a cassette player, or similar Smooth rubber roller, about 1/2" in diameter and 1/4" wide

[L] Drive wheel of any lightweight material, with a diameter slightly greater than the motor casing Between 1½" and 15/8" is good. An old VCR might have a suitable pulley, or try the disc that holds the control rods in a servomotor. You can also use a wheel from a toy, or any other right-sized plastic disc.

Rubber band

[M] Epoxy

[N] White glue

TOOLS

[O] Soldering equipment Iron, stand, solder, and solder-sucker, desoldering bulb, or braid

Dremel tool with grinding wheel, cut-off wheel, and router bits

"Third hand" tool with alligator clips Two are ideal

[P] Needlenose or long-nose pliers

Wire cutters

[Q] Hobby knife

Medium-grade sandpaper or metal file

Ruler

[R] Poster putty or tape

Safety glasses

MAKE IT.

BUILD YOUR BEAMBOTS

START Time: **A Day** Complexity: **Medium Low**

1. BUILD THE SOLARENGINE CONTROL CIRCUITS

We'll be freeforming these circuits, which means connecting components together directly, without a board. Normally I would breadboard and test my circuits before soldering, but this one is so simple and has so few parts that we can live dangerously. Parts are easily desoldered and resoldered if there's a problem.

1a. Face the two transistors up with their pins toward each other. Solder the base pin (middle) of the 3904 transistor to the collector pin of the 3906 (the right pin, as you read the printing).

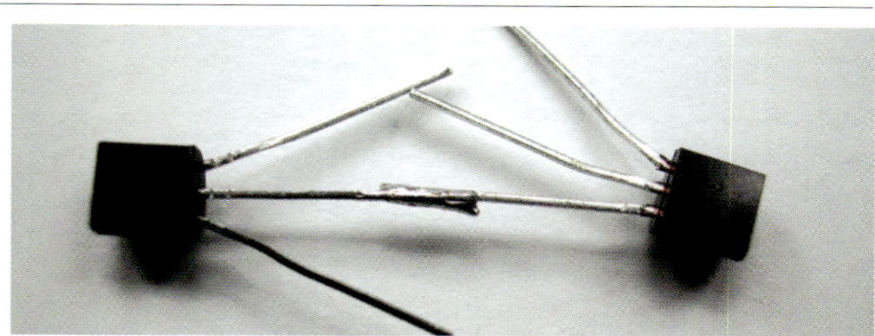

1b. Use needlenose pliers to gently bend the 3904 emitter pin (left) 90 degrees to the side and its collector (right) 90 degrees up. Bend the 3906 base pin (middle) 90 degrees up and its emitter (left) 90 degrees to the side. Solder the 2.2kΩ resistor from the 3904 collector to the 3906 base.

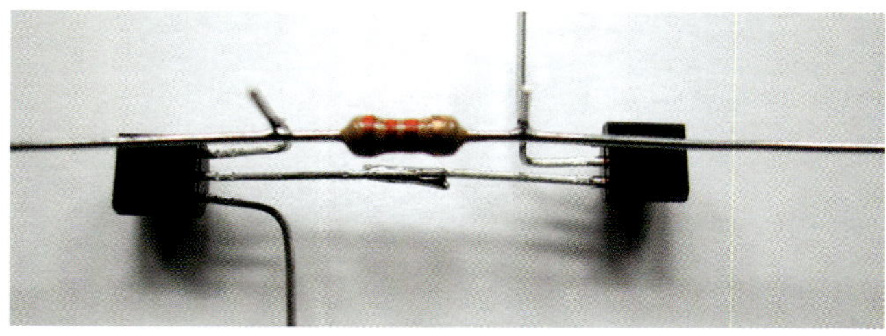

1c. Trim excess lead length from previous step. Place the 1381 voltage trigger to the right of the 3906, facing the same way. Solder its Pin 3 (right) to the 3904 emitter and its Pin 1 (left) to the 3906 collector. Finally, arc its Pin 2 (middle) around and solder it to the 3906 emitter (left). There's your basic circuit, ready for motor and power!

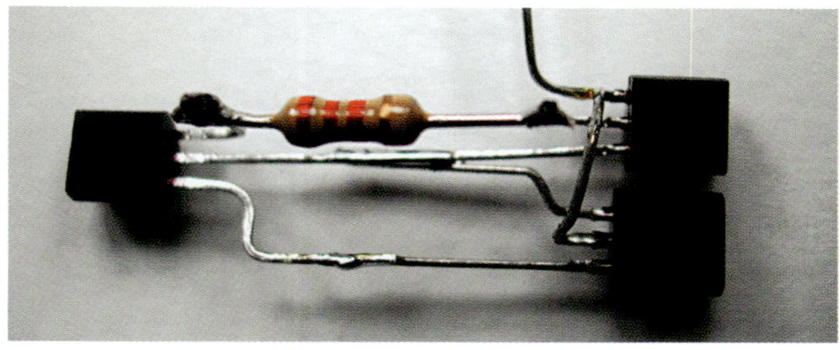

1d. If you're making both BEAMbots, build a second Solarengine circuit by repeating steps 1a-1c above. From here, you can continue on to step 2 to build a Trimet, or jump ahead to step 3 and build a Solarroller.

2. MAKE A TRIMET

2a. Prepare the motor by removing any mounting tabs with a Dremel grinding wheel. Then use sandpaper or a metal file to scuff the drive-shaft side of the case until you're down to the shiny metal underneath. Really scuff it up good; you'll be soldering capacitors directly to the case, and they'll need to hold as the Trimet drags and bumps around.

2b. Clip the negative/cathode leads on the three 4700µF capacitors so there's just enough wire to solder them to the motor casing. Bend the positive/anode leads up, making sure they comfortably clear the casing. Find 3 equidistant points at the perimeter of the motor, and solder the 3 cathodes to these points so that the capacitors form an equilateral triangle radiating out from the motor's center. Use generous gobs of solder, and use poster putty or tape to hold the caps in place while you solder.

2c. Center the circuit assembly over the motor, and solder a scrap lead from the 3904 emitter to the motor casing. This grounds the circuit, while also attaching it to the motor. For optimal balance, bend this connecting wire at 90 degrees, and try to position the circuit in the middle of the motor.

2d. The motor case is our circuit's ground (-); now let's work on the power (+) side. Take a small paper clip and bend it into a ring with the same diameter as the motor. (Conveniently, Walkman motors are the size of a quarter, so you can use one as a form to bend the clip around.) When you have a decent circle, solder it together.

2e. Bend and trim the capacitor anode leads evenly, so that they extend just above the control circuit. Solder the "power ring" to the ends of the 3 leads, preserving the equilateral symmetry.

2f. If you can bend the 3906 emitter lead to reach the paper-clip ring, do so, and solder it on. Otherwise, connect it with a short piece of wire or scrap component lead.

2g. Now, the solar cell. If yours has pre-tinned pads but no wires (most small cells come this way), start by soldering the 2 wires onto it — but be careful, because solar cells are fragile. Then solder the positive/red wire to the ring and the negative/black wire to the motor casing. Make the wires long enough so you can still work on the circuit, but short enough so they'll stow neatly underneath when you finally glue the solar cell down onto the ring.

2h. Connect the motor. Solder the negative/black motor wire to the point where the 2.2kΩ resistor meets the 3904 collector. Solder the positive/red motor wire to the paper-clip ring.

Now, place the solar cell on top of the Symet and shine a light on it, or put it in the sun. After 10 seconds or so, it should fire and scoot along, or spin around if you're holding it by the driveshaft underneath. If so, congratulations — you're the proud parent of a BEAMbot! You can go ahead and glue the solar cell onto the paper-clip ring. Or, if the cell stays in place without glue, leave it that way so that people can peek under the hood.

3. MAKE A SOLARROLLER

Solarroller builders have used all sorts of materials, from Lego bricks to soldered paper clips to computer mouse cases. This popular approach relies on parts from an old cassette player and VCR. Your mileage may vary, depending on the parts that you use for the body and drivetrain.

3a. Cut the arms on the 2 pinch rollers with a Dremel and cut-off wheel, so that they make full, flat contact against the motor casing. The Solarroller will stand on the triangular base that's formed by these 2 idler wheels and the larger drive wheel that will go onto the motor's drive shaft.

3b. Prepare your drive wheel. First, check that it will fit on the motor's drive shaft. (The hole in the hub of the disc I used was too small, so I reamed it out using a Dremel router bit.) Then glue a rubber band around the outside of the wheel, to improve traction. Cut the band, smear a thin layer of glue onto one side, and when it gets tacky, carefully roll the wheel over this "tire" until it comes full circle. Let the join overlap, then use a hobby knife to cut away excess rubber and make sure the ends are perfectly joined.

3c. Epoxy the 2 idler wheel arms into position on the motor casing, then fit the drive wheel onto the motor shaft without gluing it (use poster putty to hold it on, if needed). It is critical that all 3 wheels run parallel to each other and make full contact with flat ground when the Solarroller is standing. If you're using the Solarbotics motor, you can affix the larger roller arm to the motor's large mounting tab, pointing toward what will be the front, and leave the two other mounting tabs and holes pointing up on top, for attaching the circuit and solar panel.

3d. Cut about 4" of wire from a large paper clip and fashion it into a U shape. For the Solarbotics motor, it can be just wide enough to run between the two upper mounting holes. Trim the remaining piece of paper-clip wire and solder it across the U as a cross-brace, about ¾" from the open end.

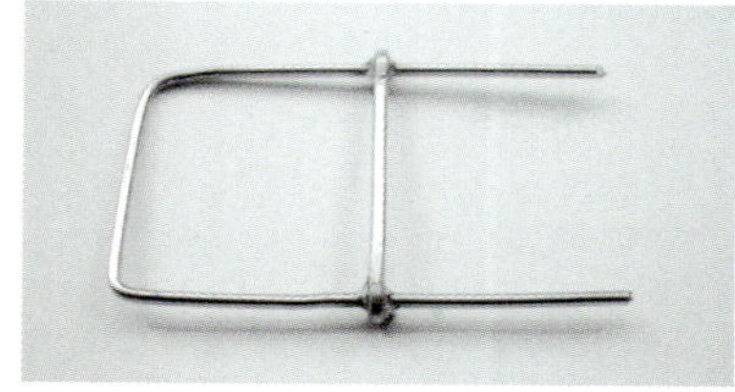

3e. Epoxy a capacitor directly to the motor casing, running horizontally, on the side opposite the drive wheel. The leads should point backward, with the cathode (-) closer to the motor.

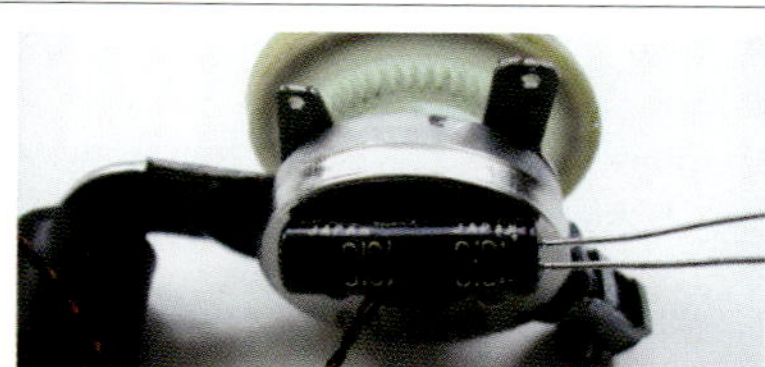

3f. Solder (or epoxy) the paper-clip frame atop the motor casing, using the two mounting holes if present. Since we didn't glue the drive wheel on yet, you can remove it to access the top of the motor. For extra sturdiness, you can position the frame so the cross-brace rests on the capacitor, and epoxy the brace onto the cap. Glue on the drive wheel.

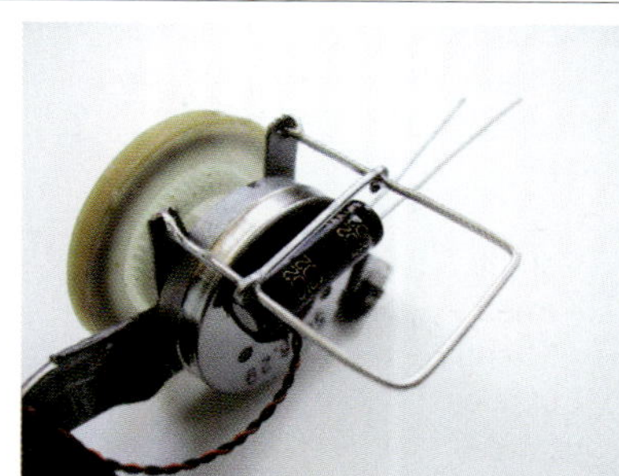

3g. Position the Solarengine circuit underneath the paper-clip frame, next to the cap, on the side opposite the motor. Solder the 3906 emitter (left) pin to the positive/anode lead of the capacitor. The connection should be short enough so that the cap holds that end of the circuit up in the air.

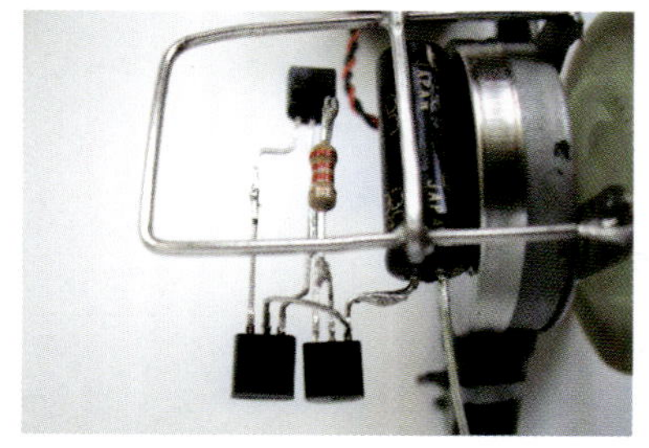

3h. Turn the Solarengine upside down and solder a scrap component lead to the 3904 emitter pin at the point where it attaches to the 1381 trigger's Pin 3. Bend the capacitor's negative/cathode lead around the undercarriage side of the cap's barrel, and solder it to the lead you just connected to the 3904. This will anchor the other end of the circuit.

Transistor Action

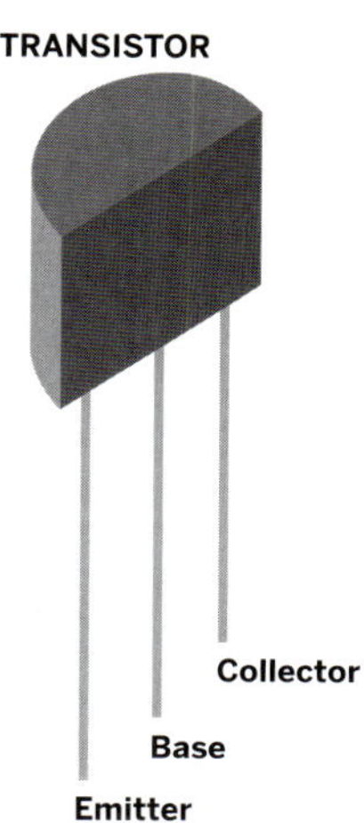

Bipolar transistors can act as switches, connecting parts of a circuit just like a mechanical switch would. In an NPN transistor, applying a voltage with the positive side to the base and the negative side to the emitter allows current to flow from emitter to collector. A PNP transistor goes the opposite way; running a negative voltage across from the base to the emitter allows current to flow from emitter to collector.

3i. If your solar cell doesn't have wires, attach some to the pads marked (+) and (-). The wires only need to be long enough to reach the pins on the capacitor. Thread the solar cell's wires through the frame and epoxy the cell to the top. When the epoxy is set, solder the solar cell's positive to the cap's positive/anode and the cell's negative to the cap's negative/cathode.

3j. Finally, connect the motor. Solder the positive/red motor wire onto the 3906 emitter (left) pin and the negative/black wire to the 3904 collector.

Now, put the Solarroller on a flat surface in the sun, or shine a flashlight on the cell. After a little while, the circuit will trigger, the capacitor will dump, and your Solarroller will take off for a short run. Shine, wait, and repeat.

NOW GO USE IT »

USE IT.

BEAM ME UP, SCOTTY

TROUBLESHOOTING

If your BEAMbot doesn't make you beam, carefully examine all connections, resolder anything that looks weak, and separate any components that might be touching (shorting). It's a simple circuit, so not much can go wrong besides incorrect connections or bad joins.

FURTHER HACKING IDEAS

On the Trimet, add an outer paper-clip ring. This creates a bumper that will help prevent the robot from getting stuck.

On the Solarroller, replace the regular 4700μF capacitor with a "supercap" like a 0.33F Gold Capacitor, as shown in the project photos. These capacitors can take several minutes to juice up, but they'll make your Solarroller take off like a bat outta hell.

You can easily convert an old Sony Walkman into a great Solarroller. Leave the motor, roller wheels, and pulleys in the original frame's base piece, and use it as the vehicle's chassis.

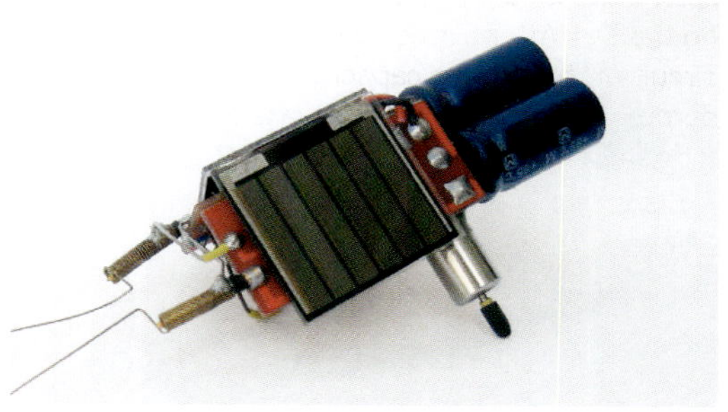

Try Andrew Miller's more efficient variant of the basic Solarengine, which is almost as easy to build. You need a different resistor, an additional capacitor, and a diode, but you can lose the 3906 transistor. Varying the value of the small cap, between 0.47μF and 47μF, lets you "program" different discharge times. (See schematic at: makezine.com/06/beambots.)

Once you have the basic ideas down, you can go crazy, improvising BEAMbots with greater storage capacity, better obstacle-avoidance strategies, or swankier, more attention-getting designs. Here are some Symet and Solarroller variations (pictured at right).

RESOURCES

There are many more hacks and variations on these two project types, as well as other applications for the Solarengine. For more information, see "Getting Started in BEAM" on page 57.

+ Schematic for Miller variant of Solarengine circuit: makezine.com/06/beambots

SOCCER-PLAYING ROBOT

By Matthew Russell

Make an autonomous robot that can chase a ping-pong ball, push it into a goal, and take other programmable actions. When you're finished, you'll have a few more wrinkles in your brain and no more fears of robot attacks. >>

Set up: p.91 **Make it:** p.92 **Use it:** p.99

Photography by Kirk von Rohr

ROBOCUP DREAMS

Robotics is one of the most multidisciplinary topics imaginable, combining artificial intelligence (AI), mechanical engineering, math, signal processing, sensor fusion, circuit design, and psychology. The resulting synergy animates applications that range from the International Space Station to kids' toys.

The most ambitious research project in robotics today may be RoboCup (robocup.org). This effort aims to develop a team of autonomous robots that will beat the human World Cup champions at soccer by the year 2050.

You can get a small taste of this complex endeavor by building your own soccer-bot, using the Lego Mindstorms Robotics Invention System. This kit includes motors, sensors, and infrared transceivers. Add a webcam and some code, and you can make an autonomous mini robot that will retrieve a ping-pong ball and bump it into a goal box. *Gooooooal!*

Matthew Russell tries to live life as a renaissance man, but hasn't made much progress since becoming enthralled by the cult of Mac.

GOAL-DIRECTED DESIGN

Showing brightly against a black field, the two soccer goals, ping-pong ball, and Lego Mindstorms robot are tracked by a webcam and simple Java software. The computer sends commands to the robot via infrared transmitter, and presto — soccer-playing bot.

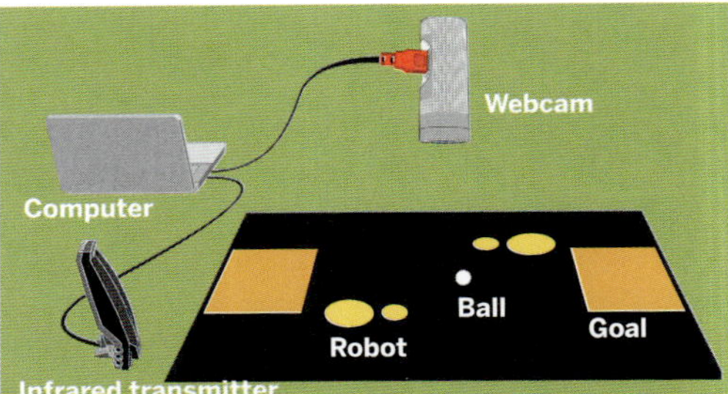

An overhead webcam above the playing field feeds an image into a computer. Software processes the image in order to identify the robot and the ball, determine the spatial relationship between the two, and calculate the robot's next moves. The software then sends the corresponding commands to the robot via infrared (IR) transmitter. The robot acts accordingly.

Our software is written in Java, which runs on most computers. Using a tool called LeJOS, we create replacement firmware for our RCX microcontroller, the big yellow Lego piece that serves as the brains of the Lego Mindstorms robotics system. The new firmware enables our Java code to control the Lego robot.

On the hardware side, we'll construct the playing field, mount the camera to the ceiling, and build the robot using Lego Mindstorms. The robot follows a simple design that resembles a tank. There's only so much room on the playing field, so the robot is small, and it follows the Lego design maxim of using as few pieces as possible to get the job done.

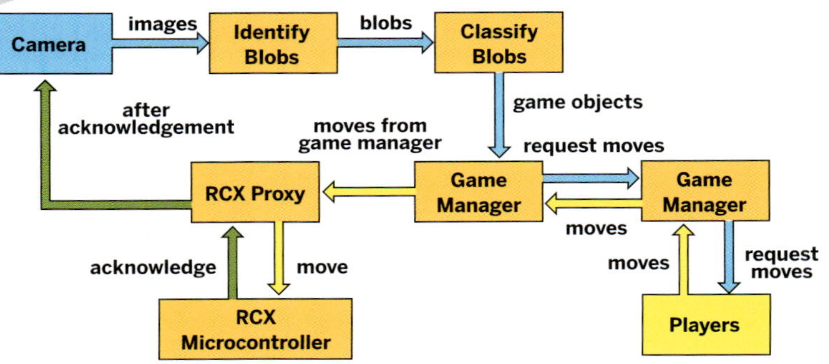

The software is where the black magic comes in. Here's the top-level loop, which continuously tracks the field and sends commands to the bot.

Illustrations by ikosy / ikosy.com

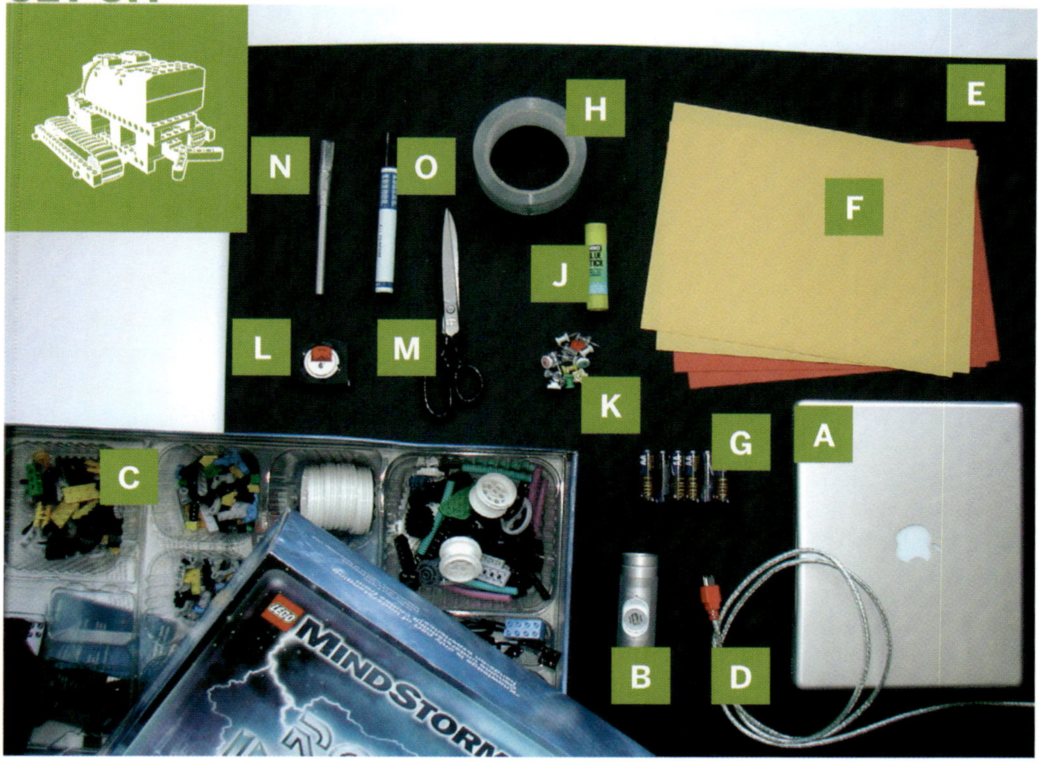

MATERIALS

[A] Computer with internet connection

[B] Apple iSight, or other webcam

[C] Lego Mindstorms Robotics Invention System (RIS) 2.0

[D] 15' FireWire cable

[E] 32"x40" black rigid-weight art boards (4), or other materials to make a flat, smooth, evenly black 3:4 ratio rectangular surface
Such as (not shown): 20"x30" white foam core boards (8), with 22"x28" black posterboards (6) to cover, or 20"x30" black foam core boards (8) — make sure it's really black, not dark gray

[F] Assorted colors of construction paper

[G] AA batteries (6)

[H] Clear packing tape

[J] Glue stick

[K] Thumbtacks

Scotch tape (not shown)

Ping-pong ball (not shown)

Velcro (optional)

Duct tape (optional)

Rubber cement (optional, if you're covering non-black foam core)

TOOLS

[L] Tape measure

[M] Scissors

[N] Utility knife

[O] Black marker

Drawing compass (not shown)

Step ladder (not shown)

MAKE IT.

BUILD YOUR LEGO SOCCER ROBOT AND STADIUM

START Time: 3-7 days Complexity: High

1. MAKE THE PLAYING FIELD

A black field sets the stage for simple, clear image analysis by reducing shadows. Goal areas are brightly colored rectangles, and a cover for the robot uses two clearly visible circles to indicate which end is front and which is back.

1a. Cut and tape together foam core, art board, or other rigid material to make a field proportional to your webcam's aspect ratio (4:3 for the 640x480 iSight and others). This lets the camera capture maximum information. If you need to cover the field with black posterboard or paper, rubber-cement it down and touch up any non-black areas with a marker.

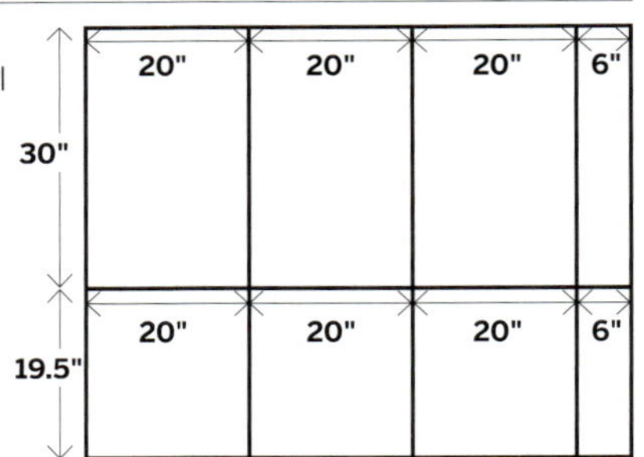

1b. Make the goal boxes. Cut two rectangles out of bright construction paper. To help the software (*see sidebar, page 97*), their width/depth ratio should be at least 1.5 or, even better, 2. Glue-stick the rectangles to opposite ends of the playing field.

1c. Make the robot cover. Pick out a different color of bright construction paper. Use a compass and scissors to cut out two circles with radii of 1" and 1.75". Glue the circles, lined up and centered, on opposite edges of a 6.5" square cut from black paper or posterboard.

Fielding Questions

The size of field needed to fill the view of a ceiling-mounted webcam will depend on the height of your ceiling.

I used eight 20"x30" sheets of white foam core, trimmed as shown, to make a 49"x66" field. Then I covered the top with 6 pieces of 22"x28" black posterboard.

You can also use 4 sheets of 32"x40" black art board to make a 60"x80" field. Just cut 4" off the width of 2 boards.

Depending on the color of your floor, it may help to put a white border around the field, by tucking paper under the edges.

To make your field easier to store, cut it back down along the edges of the original pieces, and attach Velcro across the seams underneath so you can piece it back together.

2. INSTALL THE CAMERA

You probably don't want to drill holes in your nice ceiling to mount a webcam, so here's an easy camera bracket that combines the two greatest prototyping materials ever: Lego and tape.

2a. Figure out how to snap webcam images from your computer. I used QuickTime to control an Apple iSight. If you're not using an iSight, use another webcam that's supported by Java Media Framework (JMF) or other open source drivers. Setup details can be frustrating, but they're just one of those pains that you sometimes have to wade through.

2b. Connect 2 pairs of 6-unit and 10-unit beams.

2c. Add 2 half-bushes threaded over a 10-unit axle.

2d. Snap beams crossways on top of four 2x10 plates. Thread two 12-unit beams upright over the axle, and thread a second, 8-unit axle just above. Secure beams with 4 bushes.

2e. Thread an 8-unit axle and bushes through the 8 holes up in the beams. Place the FireWire cable plug on top, and secure it firmly in place with another axle and bushes.

2f. Thread the FireWire cable between the beams along the plates. Finish off by placing two 2x8 grey plates with holes over the low ends of the beams, and plug the camera to the cable.

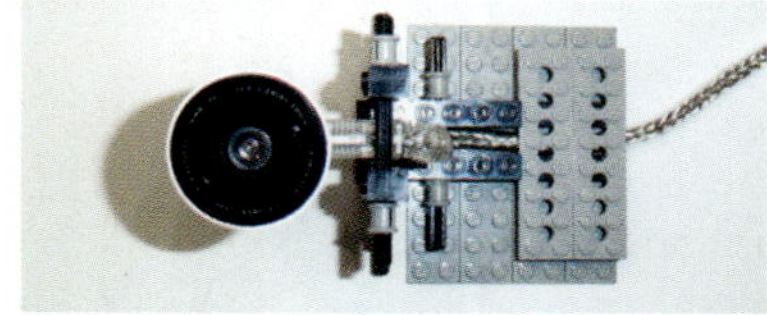

2g. Unplug the camera and secure the bracket to the ceiling with long strips of tape, reinforced with thumbtacks so it won't peel away. Run the cable along the ceiling toward the computer, and tape it up. When you're confident your bracket is secure, connect your camera and point it directly downward.

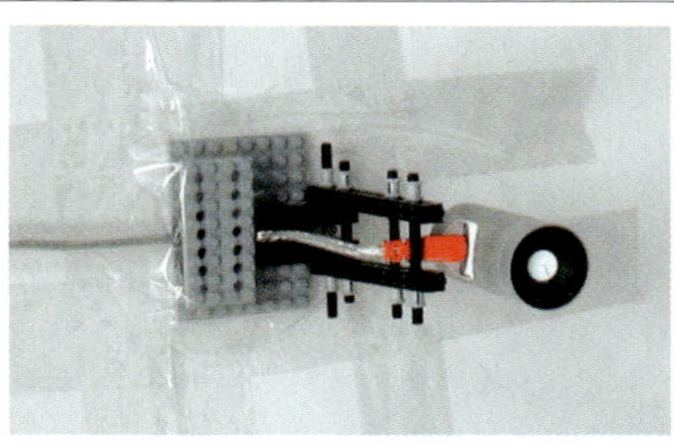

NOTE: In the legosoccer **code bundle at makezine.com/06/legosoccer, I've included a Java script named** webcampreview.sh **that uses QuickTime for previewing with the webcam. Check the *ReadMe* file for details.**

TIP: When you remove your bracket from the ceiling, use adhesive remover to clean things up.

2h. Bring up the preview pane in your webcam software. Position the field to fill the camera's field of view.

+ For advice on balancing your camera's focal length and distance, see makezine.com/06/legosoccer/focal.

3. ASSEMBLE THE ROBOT

3a. Slide 3 bushes, a wheel sprocket, and a fourth bush onto a 10-unit axle. Push them together so the fourth bush sits 1.5 bush-lengths from the end.

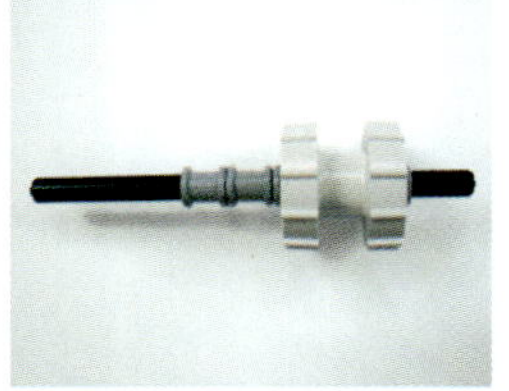

3b. Slide a bush, a wheel sprocket, a 16-tooth gear, and a 24-tooth gear onto an 8-unit axle. Push together so that the smaller gear fits inside the wheel sprocket and holds it in place.

3c. Take a 16-unit beam with the studs facing upwards. Slide the 8-unit and 10-unit axle assemblies through either end of the beam, sprocket side first, one hole in from the ends. Secure them in place with 2 half-bushes, and stretch a tank tread around the wheel sprockets.

3d. Thread a 4-unit beam and a 12-unit beam, studs pointing up, onto the other sides of the axles as shown, with the 4-unit beam on the geared axle. Secure with half-bushes.

3e. Make a mirror image of the track you just made, for the other side of the chassis. Then take four 2x8 green plates and assemble them into a square. Snap the tank tracks onto the square, with the beams staggered by 1 unit, as shown.

3f. For the robot's motor block, take two 1x2 plates with extended rails, and sandwich one corner of a 2x4 black brick, with the rails pointing outward. Place a plain 1x2 grey plate next to each, to level out the surface.

3g. Sandwich the 1x2 plates with a grey 2x2 plate above and a blue 2x2 plate below. Stack 2 more grey 2x2 plates and attach them under the black brick to level out the bottom.

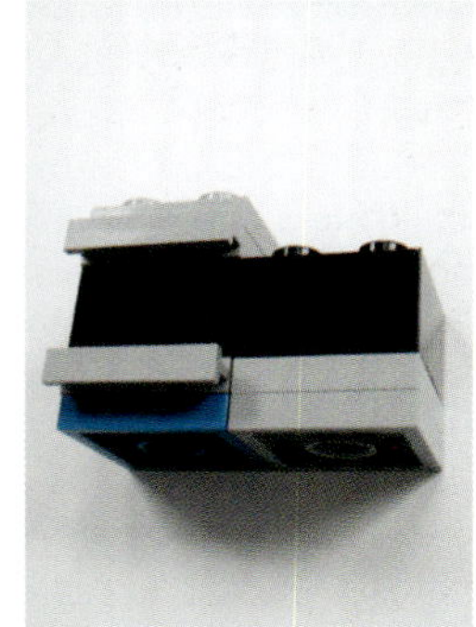

3h. Slide the rail extensions into the grooves in the side of a motor piece.

3i. Repeat steps 3f to 3h, making a mirror image, for the other motor. Place the motors side by side so the gears face outward.

3j. Join the 2 motors together with a grey perforated 2x4 plate connecting the black bricks on top. Top the inside half of the perforated plate with a 1x4 gray plate stacked on a 1x4 beam, and then reinforce the join with two parallel 2x4 yellow plates, building a platform above the motors as shown. Set a 1x2 brick on each side of the yellow plates, at the rear edge.

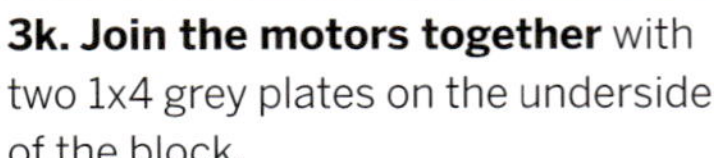

3k. Join the motors together with two 1x4 grey plates on the underside of the block.

3l. Slip a 12-tooth bevel gear over each of the 2 axles. Set the motor block on the chassis so that the gears mesh and the back of the motor block is flush with the back of the chassis.

3m. Stack 4 pairs of 2x2 blocks onto either end of two 2x6 perforated plates. Then lay these compound beams across the chassis beams in front of the motor block, one block on the very front, and the other toward the middle. These will support the deck.

3n. Place a 1x2 plate on each of two short (6") wire connectors, and place the connectors on top of the connection spots on the motor block, with wires facing outward.

3o. Make a deck with two 12-unit beams supporting a 6x10 plate, a 1x6 plate joining the beams underneath, and 2 more 1x6 plates extending along the beams above, as shown (underside).

3p. Affix the deck onto the chassis so that it overlaps the chassis' front edge by only 1 unit. The deck's 1x6 plates should flank the yellow plates. Set a perforated 2x4 plate in the center of the yellow plates.

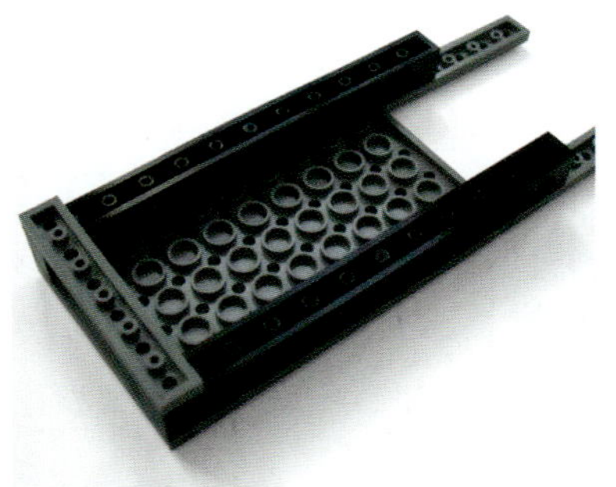

3q. Cover the hole with two 2x10 plates centered on top of the chassis, leaving the front 4 units on the 6x10 deck plate uncovered.

3r. For the robot's "grabber" arm, arrange two 1x9 lift arms so that they form a Y shape, and connect them with 1x2 and 2x2 grey plates.

3s. Insert AA batteries into the RCX unit. Place 2 pairs of stacked 1x2 gray plates on the RCX, just above the #1 and #3 connection slots. Attach the grabber arm to the grey 2x6 plate under the front of the robot, and secure the RCX on the chassis so that the front of the robot is flush. Finally, connect the cables from the motor block onto the RCX so that they're joined to opposite sides, with the cables wrapping around the outside as shown. The port (left) side motor connects to block C, and the starboard (right) motor connects to block A.

3t. Your robot is complete! Tape the cover on top, with the smaller, "head" circle facing forward, and make sure the grabber arm extends out past the cover, so the ping-pong ball will always be fully visible from above.

4. DOWNLOAD AND INSTALL THE SOFTWARE

4a. Install the Java SDK. If you've done any Java programming, you probably already have it installed. Confirm by typing `javac` and `java` into a terminal window. If not, download the latest from java.sun.com/j2se.

4b. Install LeJOS. Download the version specific to your platform from lejos.sourceforge.net. Read the documentation, which explains how to set up your Java classpath and other environment variables. The LeJOS site also has some good tutorials and a signup page for the lejos-discussion list, which is a great way to stay up-to-date with other LeJOS users (including me). After downloading LeJOS, install it and test your environment by opening up a prompt and typing in `lejos` and `lejosc`, to make sure they're in your path.

4c. Download, install, and run NetBeans (netbeans.org). Alternatively, you can run any other Java development environment, but NetBeans is what I used to develop the LegoSoccer code.

4d. NetBeans isn't aware of LeJOS by default, so we need to point it to the LeJOS libraries. On the Tools -> Library Manager menu item, create a new library item and add the four LeJOS JAR files located in your *$LEJOS_HOME/lib* directory.

4e. So that NetBeans can compile our RCX code, we also need to modify a couple of project files to make them aware of the LeJOS compiler. Follow the LeJOS FAQ instructions at makezine.com/go/lejos to make these changes. If you're not using NetBeans or don't feel like customizing it, you can also just compile the RCX microcontroller code manually.

4f. Download the `legosoccer` project source code at makezine.com/06/legosoccer.

4g. From within NetBeans, open the project files and compile each of the two project directories included, *LegoSoccer* and *RCXServer*. Be aware that by default, only the project you have set to be the "main project" will compile. So you need to select each project separately and navigate to Set Main Project for each before compiling with Build Project.

4h. Plug the Lego IR transmitter tower into your computer and position it to face onto the field.

You're done! All the pieces are in place. You've constructed a playing field for less than the cost of a few refreshments at a stadium. Your ceiling-cam is mounted and ready to provide live coverage. And a lightweight and agile player is waiting on the sidelines screaming, "Coach, put me in!" Now, let's pull out the ping-pong ball and get ready to rumble!

NOW GO USE IT »

LegoSoccer Software

As our software processes the playing field image, it performs three tasks: it locates the objects against the background, it identifies them and the directions they're pointing, and it figures out how the robot should move in order to get where it needs to go. Here's how the software accomplishes these steps.

Identifying Blobs in an Image

Our brains automatically translate what we see into a 3D map of the objects around us, but this is a complex process, as any AI researcher can tell you. To make it easier, we set up a black background and brightly colored objects. This lets our software scan the image until it finds a bright pixel. When it finds one, it searches the immediate area for more pixels that are above the brightness threshold, thereby discovering any large blobs of contiguous brightness. Then it continues on to the rest of the image. This is called a *depth-first search* because the program explores as far along each branch as possible before moving on.

Classifying Blobs

Now we want to know which color-blobs are the goals, which are the robot's head and tail, and which is the ball. We do this by looking at their colors and shapes. Our blob-classifying routine first determines the *bounding box* for each blob, which is the rectangular area that fully contains the blob. If the width of the bounding box equals its height, then we probably have a circle or ball, rather than one of the long rectangular goal areas. As a double-check, we then compare the area of the bounding box with the area of the blob inside. If the blob area is significantly smaller than the box, the blob is probably round, and the bounding box area includes corners that the blob doesn't fill. With a noisy source image, both of these simple tests can be fooled by a square-ish rectangle — which is why the goal box rectangles need to have width/depth ratios of at least 1.5.

Getting Where You Wanna Go

Now we need to tell the robot what to do. First we determine the robot's location as the midway point between the two circle's centers, and its heading as the angle formed by the two centers and the x-axis:

```
slope = (y.tail - y.head) / (x.tail - x.head)
```

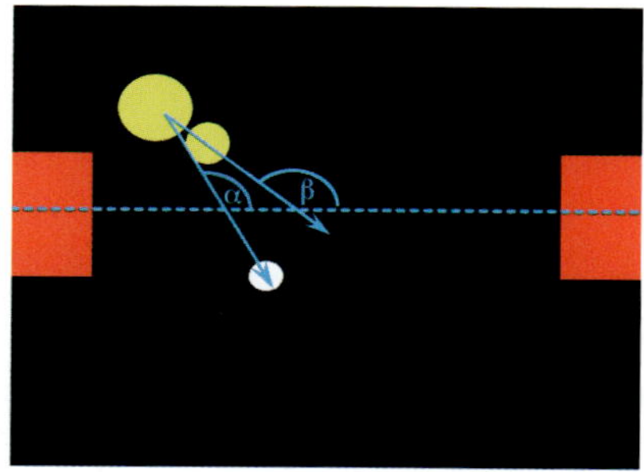

The robot's angle to the ball (α) is the arctangent of the slope between robot and ball. The difference between this angle (α) and the robot's current heading (β) determines how far it should turn.

Using calculations like this, you can get the robot to hit the ball into the goal, avoid collisions, and do other tricks. The action isn't fast-paced, since it takes time to pass the infrared and process the image. But we're stretching the limits of Lego here.

Our software defines a Java class with methods for each of these low-level behaviors. These are called, in turn, by simple shell scripts you can invoke from the command line:

`gotoball.sh`, `gotogoal.sh`, **and** `goalshot.sh`

USE IT.

LIGHTS, CAMERA, ROBOT

LEARN TO LET GO

You can remote-control your robot manually from the computer or run it on automatic using our software. But, this being the real world, performance can be imperfect either way. Like, when your robot receives the instruction to turn 90 degrees, it may turn 80, or 100. Be prepared for some troubleshooting. If a hardware glitch knocks out the IR tower, you'll need to restart, and if the IR tower loses contact when the robot is facing away, set up a shiny, reflective surface on the opposite side. If the ball keeps rolling away, put a dent in it and let it slide.

REMOTE-CONTROL PLAY

1. Place a ping-pong ball on the field. (For best results, scrub off any printing with a cleaning pad.)
2. In the `legosoccer` source code, run the `remotecontrol.sh` script. The GUI controls should soon appear on your screen, or you can use the keyboard shortcuts: W=forward, A=left, S=right, Z=backward.
3. Remote-control your robot, trying to put a ball in the goal. Remember: "Slow is fast, and fast is slow." In other words, take your time, plan your moves, and you'll get a feel for how to go where you want.
4. With two robots, you can duke it out with a `legosoccer`-enabled friend. Just make sure you assign unique IDs to your robots in the source code. Otherwise, the robots will receive and process the same commands, and act like copycats.

AUTONOMOUS PLAY

Getting your robot to work autonomously will likely require some environment-tweaking, including adjusting the lighting and trying different colors for the paper dots and rectangles. Also, make sure to keep the playing field clean and uniform; scratches, glue spots, and other glitches can confuse the software. Check the *ReadMe* files in the `legosoccer` source code for some tips on easing the frustrations.

1. Run the script `servertest.sh` from a terminal window, and follow any diagnostic advice it spits out. If all goes well, you'll see a snapshot of the field (below) with bounding boxes around all of the objects. Use this to center the field in the camera's field of view.

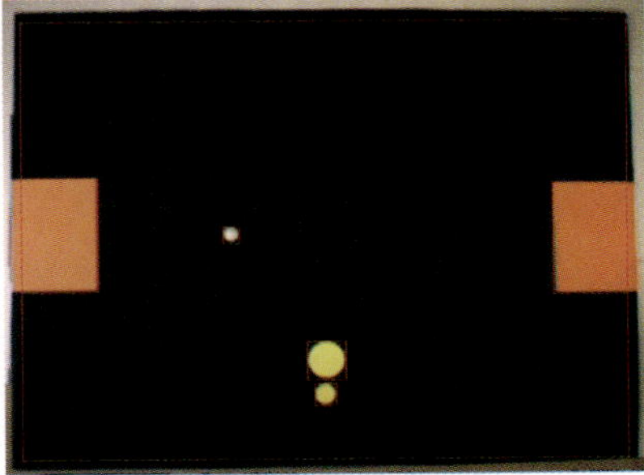

2. Run the script `gotogoal.sh`. You'll see the field again, some more diagnostics, and if all goes well, your player should turn and head into a goal box. Place your player anywhere on the field, run `gotogoal.sh`, and watch it navigate over to the goal.
3. In addition to `gotogoal.sh`, there are two more action scripts: `gotoball.sh` makes the robot hit the ball from wherever it's standing, and `goalshot.sh` makes it line up behind the ball, facing the goal, then move forward and hit it in. He shoots, he scores!

You can add as many other behavior scripts as you can write. You'll have to pay some coding dues, but the voodoo magic you'll see as a result is worth it.

RESOURCES

Core Lego Mindstorms Programming, by Brian Bagnall — Excellent introduction to Java programming, Lego Mindstorms, and LeJOS, by the creator of LeJOS.

Vehicles, by Valentino Braitenberg — Classic, thought-provoking text on designing low-level robot behavior.

BUILDING TENSEGRI[TY] MODELS

By William Gurstelle

Make a "needle tower" sculpture from dowels a[nd] elastic cord that seems [to] defy the laws of physics

Set up: p.106 Make it: p.107 Use it

HOW HIGH CAN YOU GO?

One of the most striking sculptures on display at the Smithsonian's Hirshhorn Museum in Washington, D.C., is the soaring, 60-foot-tall *Needle Tower*. It's made of a latticework of aluminum tubes held together by crossed stainless steel wires under tension.

Sculptor Kenneth Snelson's creation is a fragile-looking construction. Common sense tells you it ought to fall over or implode. Yet it's been standing strong and upright since 1968, unperturbed by storms and wind or even the occasional renegade urban climber.

The rods that make up the bulk of the tower float in midair, held together only by wires that are not guyed or anchored to the ground; they are not attached to anything but the compression rods themselves. It forces you to think: just what is pushing on what? How does it stand?

William Gurstelle is a frequent MAKE contributor. His new book, *Adventures from the Technology Underground*, was published in January.

MEET THE TENDULE

Tensegrity describes structures composed of three or more rods held in place by a network of tension wires. No rod may touch another rod.

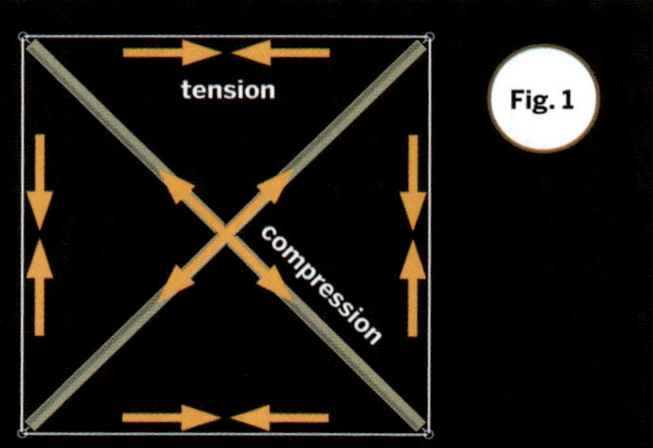

In its most basic form, a tensegrity structure is simply a construction made from two crossed rods, held in place with a tightened cord around its perimeter (Figure 1). It's easy to see that if the cord is in tension, then the rods are under compression. But this kite shape is actually a quasi-tensegrity, because the crossed rods touch each other in the middle. In a true tensegrity, no compression members make contact.

It's a only a small step from the kite frame to the simplest true tensegrity structure, a basic, versatile, and elegant structure which I call a tendule (Figure 2).

Tendules have three compression members that form a stable and (unlike the kite shape) three-dimensional building block. This is the basic building unit of Snelson's *Needle Tower* and a host of other tensegrity structures.

The three essential components of a tensegrity tendule are as follows:

1. The compression members (wooden dowels)

2. The tension members (stretchy cords), which consist of:
A. The cord triangle loops: The two tension members in every tendule that connect the three compression members in the horizontal plane.
B. The vertical cords: The three tension members in every tendule that connect the top of each compression member to a corresponding bottom in an adjacent compression member.

3. The tension-compression interface (eye screws)

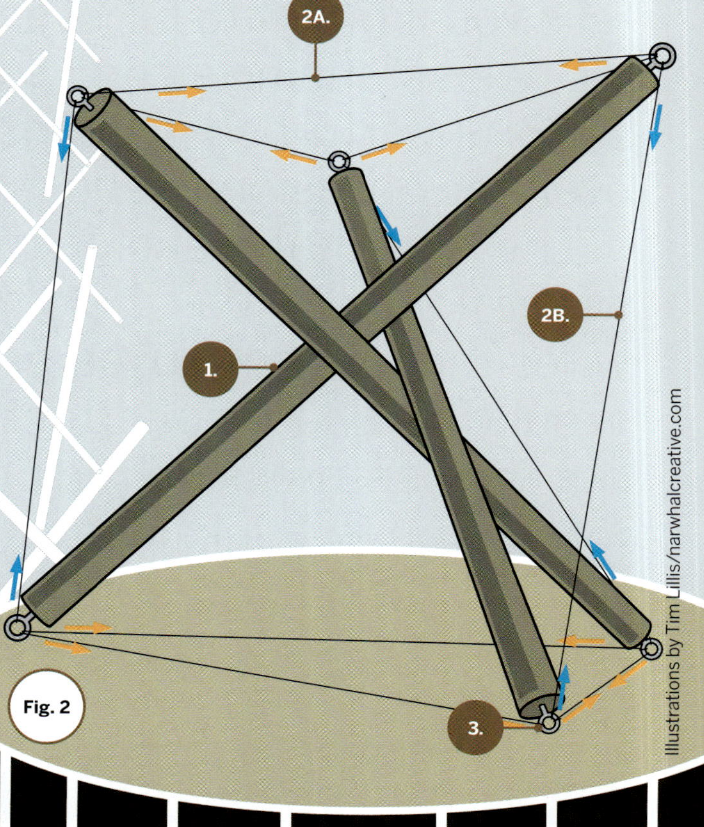

Illustrations by Tim Lillis/narwhalcreative.com

A SHORT HISTORY OF FLOATING COMPRESSION STRUCTURES

Photography detail courtesy of the Hirshhorn Museum and Sculpture Garden, Smithsonian Institution, Gift of Joseph H. Hirshhorn, 1974. Image by Ricardo Blanc

The key to understanding the *Needle Tower* and a number of other Snelson works is a principle popularly referred to as *tensegrity*, a combination of *tension* and *integrity*. This is a neologism coined by Buckminster Fuller, a man who loved to create new words by truncating and fusing others: *dymaxion* (*dynamic maximum tension*), *synergetics* (*synergetic geometry*), and *sunsight* (*sun* and *first sight*).

Origins of Tensegrity

Some writers claim the notion of tensegrity was first put forward by Snelson in 1948, when he described a technique he used in building artistic constructions as "floating compression." Snelson shared his ideas with Fuller, who quickly saw both novelty and practicality in the idea. An expert at providing interesting names to concepts, he reworked and then renamed this generalized idea of structures held together by compression members, which are held in place by flexible wires in tension.

When Fuller coined and began popularizing the term *tensegrity* in 1961, apparently he did so without Snelson's consent. In fact, there is something of a controversy among scientists, artists, and architects as to who rightfully deserves more of the credit for devising the concept. Snelson claims quite vociferously that the famous and powerful Fuller more or less usurped his idea and denied him recognition, while Fullerites scoff at the idea that anybody but Bucky deserves credit.

Yet others maintain that the idea predates both of them, and belongs either to an obscure Latvian artist named Karl Loganson, who some say developed the idea in the early 1920s, or to French architect David Georges Emmerich, who began exploring and refining the basic building structures of tensegrity in the 1950s.

Tensegrity's true provenance is difficult to ascertain given the number of competing claims and the documentation on the subject. Many, if not all, of these erstwhile creators likely had a hand in developing and extending the concept. Almost certainly, each advanced the idea in important ways. Emmerich, Fuller, and Snelson each filed patent claims on differing aspects of the technology in the 1960s.

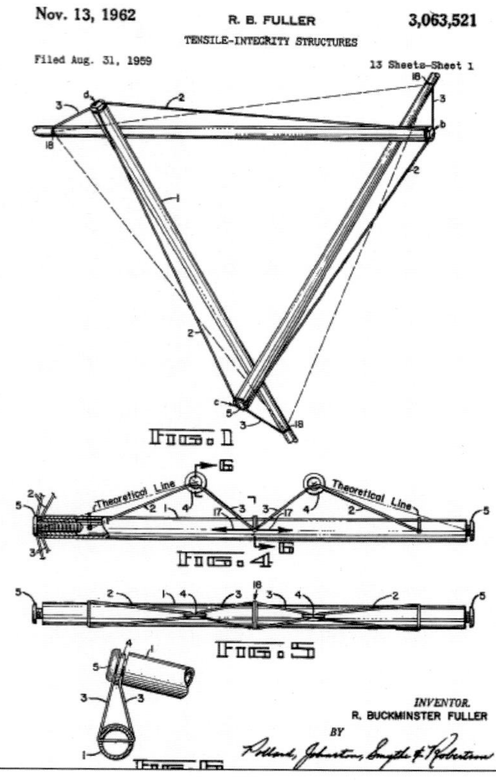

Illustration from the U.S. Patent & Trademark Office website

Now, it appears that each man had a much different opinion on the highest and best use of this idea. Fuller's primary interest in the concept was its application toward engineering — specifically, the spherical and dome-like enclosures (i.e., geodesic structures) for which he became famous. Emmerich, the architect, saw tensegrity's value residing primarily in its utility as a building construction technique. And Kenneth Snelson, the artist, primarily sees tensegrity's value in an aesthetic sculptural context.

Irrespective of who actually first developed the concepts and the principles behind the building technique, tensegrity structures are fun to make and interesting to look at.

Toward a Floating Tower

Ever want to make something that is cool simply for its own sake? The English might term such a project as "folly" — an eccentric sort of activity without much practical value, but valuable because of its interesting and out-of-the-ordinary nature. It's the sort of unusual, impractical, but enthralling project that makers seek out. If you're a regular reader of this magazine, you know what I'm talking about. This is one of those projects.

Be aware, the project that follows can lead to a pleasant sort of addiction. Once you master the techniques shown here, you may find yourself building a great number of tensegrity structures. Also, be forewarned that the first time you build one, it will be frustrating. Cords will break, the modules that make up the structure may be uneven and tilted, the towers will lean or list, or worse, come crashing down in a heap. Eventually, the technique will make sense. The materials are somewhat forgiving, and misalignments can often be corrected by gentle massage and your force of concentration.

A Few Background Notes

I first came across the idea of tensegrity in a 1996 edition of *The Physics Teacher* magazine. The author of that article, a University of Michigan faculty member, described his encounter with Snelson's *Needle Tower* at the Hirshhorn Museum as an experience that left him puzzled and intrigued. The author provided directions and sketches to enable the magazine's readers to replicate on a smaller scale the fragile-looking, semi-rigid column that seems to float on air.

When I tried to make my own column, I was less than successful. I followed the instructions closely, but it was apparent that the technique described in the article required either the manual dexterity of a Flying Karamazov chainsaw juggler or a prehensile tail to act as a third hand. Lacking both, the structure I tried to build invariably collapsed, sometimes slowly, sometimes quite spectacularly, and often just on the brink of success.

But, perseverance is the mother of good luck, and I eventually came up with an easy-to-replicate modeling technique that works quite well, even for people limited to just ten fingers.

Building Tensegrity Structures

In Fuller's 1962 U.S. patent application, tensegrity is described as "a plurality of discontinuous compression columns arranged in groups of three nonconjunctive columns connected by tension elements forming tension triangles." Huh? No wonder tensegrity sculptures seem hard to make.

Buckminster Fuller built this geodesic dome in 1962. He developed some of his ideas about tensegrity after meeting a young Kenneth Snelson in the late 1940s. A fascinating letter written by Snelson and printed in the International Journal of Space Structures in 1990 is reprinted here: grunch.net/snelson/rmoto.html.

Photograph courtesy, The Estate of R. Buckminster Fuller

Happily, a lot of other people, from artists to scientists, have developed their own definitions:

> *Tensegrity is a structure composed of three or more rods held in place by a network of tension wires. No rod may touch another rod. Rods may be supported in place solely by the tension wires to form a firm, prestressed, and semi-rigid structural unit.*

This description is somewhat more user friendly. It provides a set of conditions that can be used to determine whether a structure uses tensegrity construction techniques or not. But after you work with it a while, tensegrity becomes easy to identify, even if it remains difficult to describe in words. To paraphrase Supreme Court Justice Potter Stewart's 1964 quote, "I shall not attempt further to define this kind of material, but I know it when I see it."

Floating Compression Modeling

Tensegrity models can be made from inexpensive materials such as hardwood dowels and elastic cord, or even just rubber bands and pencils. More elaborate models occasionally are made using brass rods or steel shafts.

All tensegrity structures described in this article are built up from basic building blocks called tension modules (which, to be Fullerian about it, I call *tendules*). A tendule is composed of three rods that support compressive loads and five triangulated wires or bands that handle tension loads. When arranged in a specific style of latticework, the rods and wires form stable building blocks that can be turned into towers, bridges, and other, even more elaborate space-enclosing structures.

To take on larger shapes, the tendules rely on a building principle called *floating compression*. None of the rods touch one another, and loads are transmitted from rod to rod through a matrix of tensioned wires. Because the wires are very thin and the rods are relatively thick and widely separated in space, the overall visual effect is that of unsupported struts or poles floating in air — a most unusual and novel spatial aesthetic.

Building a Tensegrity Tower

The instructions in the following section detail the materials and techniques used in building a model of Snelson's *Needle Tower*. The model is four tiers or stories high, although fewer or more tiers can be incorporated according to the wishes of the builder. The original *Needle Tower* at the Hirshhorn Museum has more than 15 tiers.

SET UP.

MATERIALS

5m-long spool of 0.8mm to 1.0mm diameter clear plastic stretch cord Pepperel Braiding Company's Stretch Magic cord worked for me. Available at most crafts and beading stores.

36" long, 5⁄16" diameter wooden dowels (3) Other diameters will work satisfactorily, but I find 5⁄16" diameter to be the most aesthetically pleasing.

#214 eye screws (24)

Box of assorted medium-sized rubber bands

Masking or electrical tape (optional)

Spray paint (optional)

Wooden or tile base (optional)

TOOLS

Handsaw

Standard or locking pliers

Small regular screwdriver

Electric drill

5⁄64" drill bit

Scissors

MAKE IT.

CONSTRUCT A TENSEGRITY TOWER

START Time: A Day Complexity: Low

1. FABRICATE THE COMPRESSION MEMBERS

In tensegrity structures, the tension (pulling) stresses and the compressive (pushing) stresses are completely separated. The structure's compression loads are carried by rigid wooden dowels.

1a. Cut the wooden dowels to 8" lengths using the handsaw.

1b. Open up the end of the eye screws slightly with the pliers and a small screwdriver or needlenose pliers.

1c. Drill a 5⁄64" hole in the center of each end of each dowel, deep enough to accommodate the stem of the eye screw.

TIP: Roll a width of masking or electrical tape around each end of each dowel to prevent end splits.

1d. Insert eye screws into holes. There are 3 dowels required per story. A 4-story tower requires 12 dowels.

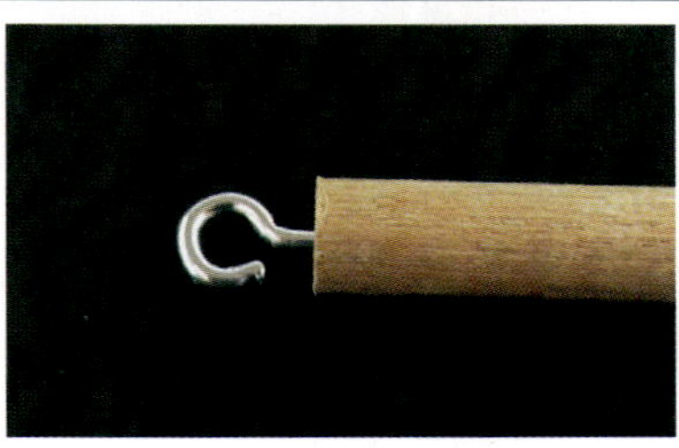

2. FABRICATE THE CORDS

Tension in the structure is provided by loops of stretchy plastic cord. The cord packaging should carry instructions on how to tie the cord into loops. If there are no instructions, try either a square knot or a double overhand knot, pulled as tight as possible.

2a. Make 8 loops from stretchy cord pieces, each piece being 15" long. Making the knot will consume an inch or so of cord, so the total length of the cord in the loop will be slightly less than 15". Because the cord is stretchy, there is some wiggle room in making the loops.

2b. Make 9 loops from cord pieces cut 8" long.

2c. Make 12 loops from cord pieces cut 10" long.

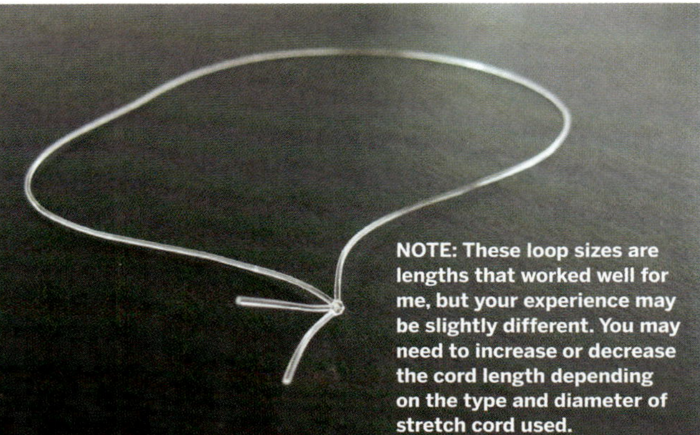

NOTE: These loop sizes are lengths that worked well for me, but your experience may be slightly different. You may need to increase or decrease the cord length depending on the type and diameter of stretch cord used.

3. CONSTRUCT THE WEAVE TRIANGLES

Tendules begin as a set of dowels arranged in groups of 3 to form preliminary weave triangles. There will be 2 left and 2 right weave triangles.

3a. Place a pencil mark at a point 3" from each end of each dowel. Align dowel pencil marks to form a left weave triangle. Tightly wrap a rubber band around the pencil marks to hold the triangle together. Repeat for the second left weave triangle.

3b. Now align the dowels to form a right weave triangle, pencil mark to pencil mark, and wrap it tightly with rubber bands. Repeat for the second right weave triangle. You now have 4 flattened weave triangles, 2 left-handed and 2 right-handed. (You can make more later, if you want to build a higher tower.)

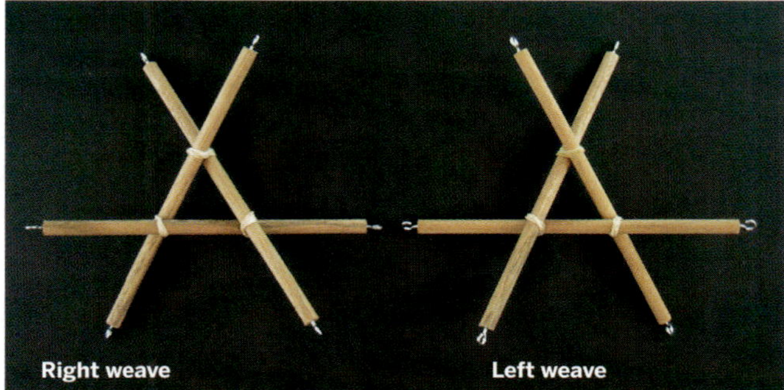

Right weave

Left weave

NOTE: The rubber bands are only temporary. You'll cut them off later.

4. FORM THE CORD TRIANGLES

The cord triangles are the triangles formed by the stretchy cords. In the photo below, weave triangles are the triangles formed by the wooden dowels, and the cord triangles are the triangles colored red and blue.

4a. Attach the 15" cord to the top set of opened eye screws on a weave triangle (as shown in blue). Then attach another 15" cord to the bottom set of eye screws on the triangle (as shown in red). Note how the top and bottom stretch cords form 2 cord triangles.

4b. The best way to form the cord triangles is to loop the cord once around the base of each eye screw. The cord should be stretched tightly but evenly in each cord triangle; that is, each third of the triangle should have roughly the same tension. You can easily check the tension by plucking each segment of the stretched cord triangle with your fingernail and listening to the resulting pitch. When the plucked pitches are all equal, the tension in each segment is equal as well.

4c. Adjust the tension by wrapping extra turns of the cord around the eye screw post until the tension in all 3 segments of each cord triangle is the same. Although the plastic cord is applying lateral tension to the weave triangle, the rubber bands should prevent it from moving. If the weave triangle shifts or is unstable, tighten the rubber bands to prevent movement.

5. ATTACH THE VERTICAL CORDS

5a. Attach one of the 10" long cord loops to each top weave triangle eye screw and to the eye screw on the bottom triangle closest to it. The vertical cord connecting the eye screws (top to bottom) will be very loose at this point. Initially, you may need to loop the vertical cord once or twice around the eye screw base to keep it from slipping off.

6. FORM THE TENDULES

Tendules are the basic tensegrity building blocks. A single tendule is a compression-tension structure that is freestanding and stable. Larger structures are made by linking one tendule to another.

6a. Use scissors to carefully cut the rubber bands. When the last band is cut, the tension within the upper and lower triangle cords will lift up the structure to form a 3D shape, while the tension in the vertical connecting cords will constrain the structure from moving too far. The result is a box-like shape. Finding the correct balance between the lateral (cord triangle) tension members and the vertical tension members aligns the compression members into semi-rigid, 3D tendules.

6b. Fluff and adjust the tendule by looping the plastic cords over the eye screws until it forms a box-like shape. The top and bottom triangle tension members are in blue, and the vertical connecting cords are in white.

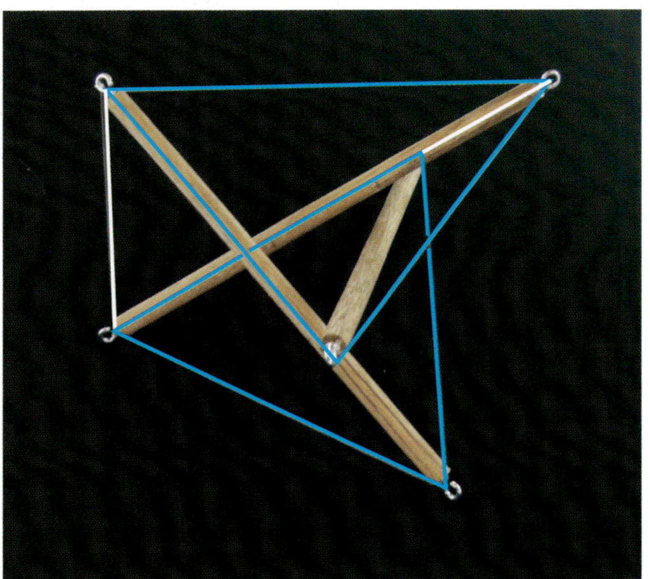

6c. If the tendule droops, loop the top and bottom triangle cords additional times around the eye screw post while simultaneously reducing the tension in the vertical connectors. I've found that the hardest part of making a tensegrity tower is learning how to properly adjust the tension ("tuning") in the plastic cord loops. Each adjustment affects all other tension members — it's a sort of mechanical sudoku puzzle for your hands — so considerable tuning is required.

6d. Start by making single loops over the eye screws. Go easy at first on the number of loops, because making too many loops too quickly will cause over-tightening in one plane of the tendule and misalignment in the others.

6e. Check the tension by plucking the cord after each eye screw loop adjustment until the tension is evenly distributed and the desired boxy shape is attained.

6f. Once you've figured out how to manipulate the tension in the triangle to form an evenly aligned tendule, make 3 more.

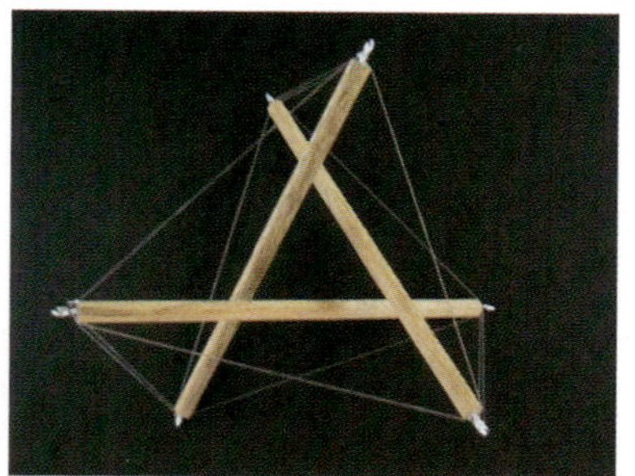

7. ERECT THE TOWER

The tensegrity tower is erected by placing one tendule on top of another, alternating left-handed and right-handed tendules as you go. Each tendule is fixed to the ones above and below by triangulating. This means that tension members (additional stretchy cords) are affixed to form stability triangles that connect stories together.

7a. Place one tendule on the base. Set a tendule of the opposite handedness on top of the base tendule. Rotate the upper tendule so that its eye screw touches the triangle cord on the base tendule halfway between eye screws. Loop the upper tendule's eye screws around the triangle cord (colored red in the photo) several times. Adjust tension in the tower cords by looping until the tension in all cords is equal.

7b. Repeat for the third and fourth tendule in the tower, alternating the handedness of the tendule each time.

7c. Once all the tendules are stacked, it is time to triangulate the tower to make it rigid. Do this by connecting the 9 remaining 8" loops to form stability triangles between each row. Complete the stability triangles by extending the 8" loop from each bottom eye screw to the bottom eye screw directly below it.

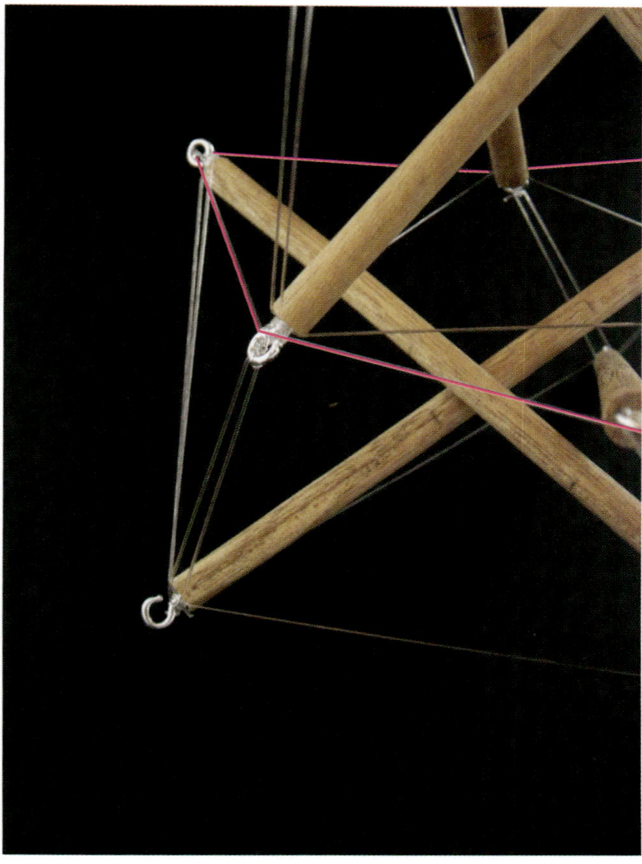

7d. The colored lines in this photo show how each stability triangle is formed between the ends of 2 compression members in the lower tendule and the end of the descending compression member in the tendule above.

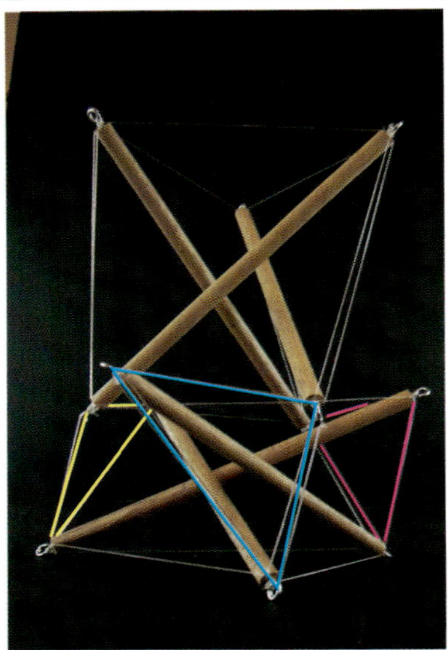

7e. Adjust all cords so that the tension in each tendule is even and the cords that connect the stacked tendules are evenly distributed. Again, test by plucking and listening. Add or subtract loops around the eye screws to even up tension in the structure.

7f. After adjustments are complete, the tensegrity tower should be stable and level.

USE IT.

TENSEGRIFY YOUR WORLD

HINTS AND TIPS:

1. Often, much fluffing and tension adjusting is required to get the tendules to stand squarely. Take as much time as necessary to get this right.

2. Once you learn the basic technique, you can experiment with unequal-length compression members and variable tensions to obtain interesting shapes. Interesting variations include: square and five-sided weaves, using tendules whose compression members are all longer or shorter than the members in the adjoining tendule, or using compression members of varying sizes in the same tendule.

Fuller vs. Snelson on the Definition of Tensegrity

"The word 'tensegrity' is an invention: a contraction of 'tensional integrity.' Tensegrity describes a structural-relationship principle in which structural shape is guaranteed by the finitely closed, comprehensively continuous, tensional behaviors of the system and not by the discontinuous and exclusively local compressional member behaviors. Tensegrity provides the ability to yield increasingly without ultimately breaking or coming asunder."

—Buckminster Fuller
(from *Synergetics*, p. 372)

"Tensegrity describes a closed structural system composed of a set of three or more elongate compression struts within a network of tension tendons, the combined parts mutually supportive in such a way that the struts do not touch one another, but press outwardly against nodal points in the tension network to form a firm, triangulated, prestressed, tension and compression unit."

—Kenneth Snelson
(kennethsnelson.net/faqs/faq.htm)

FOR FURTHER STUDY

Tensengrity is a fascinating topic and there is much material available for further research.

Books:

An Introduction to Tensegrity by Anthony Pugh, University of California Press, 1976.

Engineering a New Architecture by Tony Robbin, Yale University Press, 1996.

Websites:

Kenneth Snelson: kennethsnelson.net

Buckminster Fuller: bfi.org

A Practical Guide to Tensegrity Design (a PDF file): angelfire.com/ma4/bob_wb/tenseg.pdf

Other:

Images and Community: tensegrity.com

How to Build Tensegrity Structures out of Soda Straws: georgehart.com/virtual-polyhedra/straw-tensegrity.html

The Hirshhorn Museum: hirshhorn.si.edu

Tensegrity in a Cell: makezine.com/go/cell

OWN YOUR OWN Tower PC

The standard tower-style PC is blessedly modular, designed to be opened up and upgraded.

By Robert Bruce Thompson

Tower PCs are endlessly modifiable. Most tower PC upgrades take only a few minutes and require no more than a Phillips screwdriver. At relatively small cost, you can increase your system's performance, improve reliability, reduce noise, extend life, and add features. Here, from the recent book, *Repairing & Upgrading Your PC* (O'Reilly), are the ten most popular PC upgrades and how you make them.

A CPU Extend the useful life of your system and dramatically increase performance with a faster CPU. Check documentation to verify which CPU models are compatible with your motherboard.

- **Remove the CPU cooler fan, lift the ZIF lever to release the old CPU, and remove it from the socket. Drop the new CPU into the socket, re-latch the ZIF lever, apply thermal compound, and install a new fan.** ($50 to $150)

Photograph by Robert Bruce Thompson

B CPU Cooler Replace this to reduce noise and CPU stress. Choose a premium, quiet fan from Arctic Cooling, Thermalright, or Zalman. Verify that it's compatible with your CPU and motherboard. • **Mounting methods vary, so follow the directions included with the cooler you choose. Make sure to apply thermal compound per the instructions. ($15 to $50)**

C Case Fans Add one or more fans (or replace current fans) to improve cooling, increase system stability, and reduce noise. Choose the largest fans for which your case provides mounting locations, favoring quiet models from Panaflo, Nexus, Arctic Cooling, and others. Models with variable speed settings let you balance cooling levels against noise. • **Install the fan using the four provided screws or flexible mounting studs, and connect the power lead to a motherboard fan power header. ($5 to $15)**

D Optical Drive Replace a CD/DVD-ROM drive or CD writer with a DVD±R/RW writer for fast, high-capacity backups and DVD duping. Pick a standard (parallel) ATA model that supports at least 16X DVD±R, 4X DVD±RW, and 4X DVD±R/DL dual-layer, such as the Plextor PX-716A or PX-740A, NEC ND-3540A, or BenQ DW-1640. • **Detach cables from old drive, unscrew four screws, and remove. Reverse the process to install new drive, making sure it's jumpered the same (master or slave) as the old one. ($40 to $100)**

E Hard Drive Install a second hard drive to store audio and video files, photos, and other personal data. Choose a 7,200 rpm ATA or SATA model, such as a Seagate Barracuda 7200.9, from 40GB to 500GB, depending on your needs and budget. • **First, verify that the master/slave jumper is set correctly (required for ATA models only). Then slide the drive into a drive bay and secure it with the four screws provided. Connect the power and data cables, restart the system, and format the drive. ($50 to $375)**

F Power Supply An ATX 2.0 power supply can support additional peripherals, improve cooling, reduce noise, and increase system stability. Choose a premium, noise-reduced model with a wattage rating at least 50% higher than your current unit. • **Disconnect power cables from the motherboard and drives, then unscrew four screws to remove power supply. Reverse the process to install the new unit, making sure to reconnect all of the cables. ($50 to $125)**

G Memory Add memory to improve system performance and stability. Consider 512MB a minimum for a lightly loaded Windows XP system, and 1GB for a mainstream system. Buy only name-brand memory, such as Crucial or Kingston, and use their online configurator/advisor to find modules that will match your system. If your motherboard supports dual-channel memory operation, install modules in pairs. • **Press new modules into empty memory sockets, making sure they seat completely, and that the socket retaining arms snap into place. ($25 to $100)**

H Video Card (not shown) Replace your current video card or integrated video adapter with an AGP or PCI Express video card to improve 3D gaming performance, support a second display, or add a digital interface (DVI) for a digital-only LCD display. • **Remove the old video card (if any), align the new video card with the AGP or PCI Express slot, and press the card firmly into place until it seats fully. ($25 to $500)**

I Sound Card (not shown) Replace your current sound card or integrated audio adapter with a PCI sound card to add hardware-accelerated audio for gaming and surround-sound support. Choose a SoundBlaster Audigy-series card for the best gaming audio, or an M-Audio Revolution card for both excellent gaming support and superb sound quality. • **Remove the old sound card (if any), align the new sound card with an available PCI slot, and press the card firmly into place until it seats fully. Follow the card's installation directions carefully, noting how and when to remove the old audio drivers and install the new ones. ($50 to $150)**

J Motherboard This Big Kahuna upgrade brings your system up to current standards. Choose a motherboard that fits your case and is compatible with the CPU and memory you plan to use. If you also replace the CPU and memory, you essentially have a whole new computer at a cost of $125 to $350 and an hour's work. • **Disconnect all cables from the old motherboard, remove mounting screws, and lift it out of the case. Attach stand-offs at each location required by the new board; install CPU, CPU fan, and memory on the board; then slide it into position. Install mounting screws and reconnect all cables. ($50 to $125)**

Robert Bruce Thompson is the author or co-author of several PC hardware books from O'Reilly Media, Inc., and has been building, repairing, and tinkering with PCs for more than 25 years.

1+2+3

LED Throwies

By Graffiti Research Lab

Make and toss a bunch of these inexpensive little lights to add color to any ferromagnetic surface in your neighborhood.

You will need:

10mm diffused LED, any color(s) 20 cents each from HB Electronic Components (hebeiltd.com.cn).

1" strapping tape
One roll will make many throwies.

CR2032 3V lithium batteries
25 cents each from CheapBatteries.com.

½"x1" NdFeB disc magnet, Ni-Cu-Ni plated
25 count for $13 from Amazing Magnets (amazingmagnets.com).

Conductive epoxy (optional)
Weather-resistant alternative to tape. Available from Newark InOne (newark.com).

1. Test the LED.

Pinch the LED's leads to the battery terminals, with the longer lead (the anode) touching the positive terminal (+) of the battery, and the shorter lead (the cathode) touching negative (-). Confirm that the LED lights up.

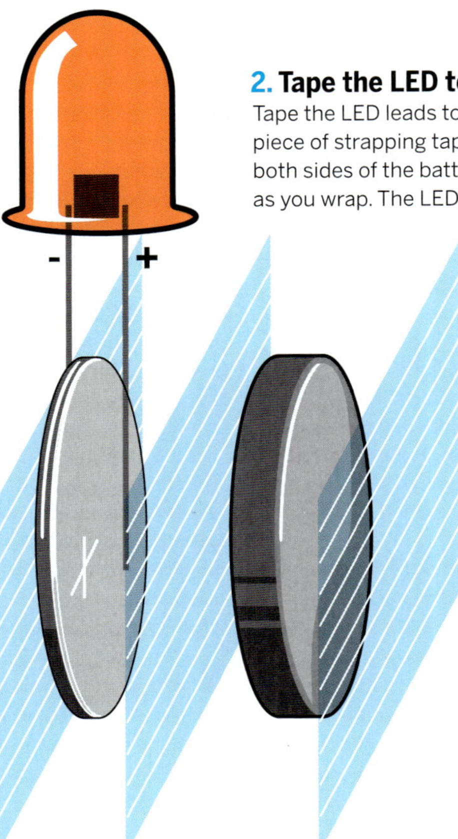

2. Tape the LED to the battery.

Tape the LED leads to the battery by cutting off a 7" piece of strapping tape and wrapping it once around both sides of the battery. Keep the tape very tight as you wrap. The LED should not flicker.

NOTE: The battery's positive contact surface extends around the sides of the battery. Don't let the LED's cathode touch the positive terminal, or you'll short the circuit.

3. Tape the magnet to the battery.

Place the magnet on the positive terminal of the battery, and continue to wrap the tape tightly until it's all done. The magnet should hold firmly to the battery. That's it — you're ready to throw (or make a few dozen more). Throw it up high and in quantity to impress your friends and city officials.

Throwies naturally chain together in your pocket, making multi-segmented throwie bugs, which will also stick to metal surfaces if they aren't too long.

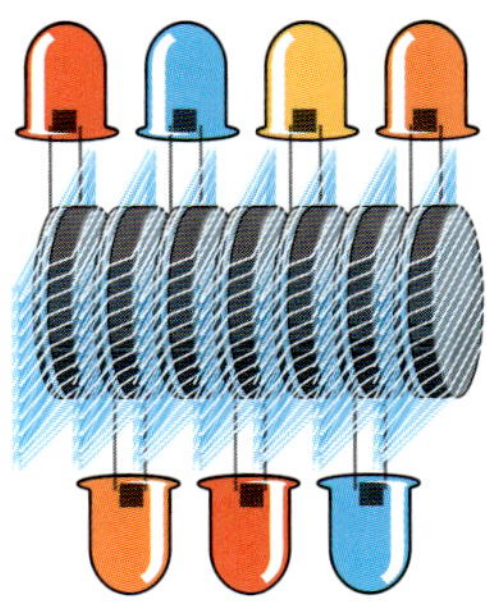

A throwie will shine for about 1-2 weeks, depending on the weather and the LED color. To get one off a ferromagnetic surface, don't pull it, or it may come apart. Instead, apply a lateral force to the magnet base, and slide it off the surface while lifting it with a fingernail or tool.

Illustrations by Kirk von Rohr

Graffiti Research Lab (graffitiresearchlab.com) is dedicated to outfitting graffiti artists with open source technologies for urban communication.

The time just seems to fly by when your unread email is piling up.

IT'S EMAIL TIME

Innocent-looking "clock" monitors the unread-message pileup in your inbox.

By Tom Igoe

I have a lot of anxiety about email. Every kilobyte in my Inbox destroys another minute of my life, but I can't stop checking it. So I decided to embody my anxiety in a device that would worry about my incoming mail for me. I've always liked clockwork mechanisms, so I made my email fetish object in the form of a clock. For each kilobyte of new mail I receive, the clock ticks relentlessly forward.

Here's the basic design I came up with. The clock itself is driven by a microcontroller, which connects to the internet and queries a program that checks my email accounts. The program reports back the number of kilobytes, and the microcontroller moves the clock forward a tick for each kilobyte. Simple!

To build this yourself, you need to know how to program a microcontroller and how to do some basic web programming. For microcontroller programming, see the Primer article in MAKE, Volume 04 (*page 158*), or my book *Physical Computing: Sensing and Controlling the Physical World with Computers*. For the web component, I used a Common Gateway Interface (CGI) script written in Perl, but you can also write CGI scripts in PHP, Python, Ruby, and other languages.

Building It

I started by looking for the simplest way to drive the clock. I carefully took the clock apart and examined the circuit board inside. There was a

Photography by Tom Igoe

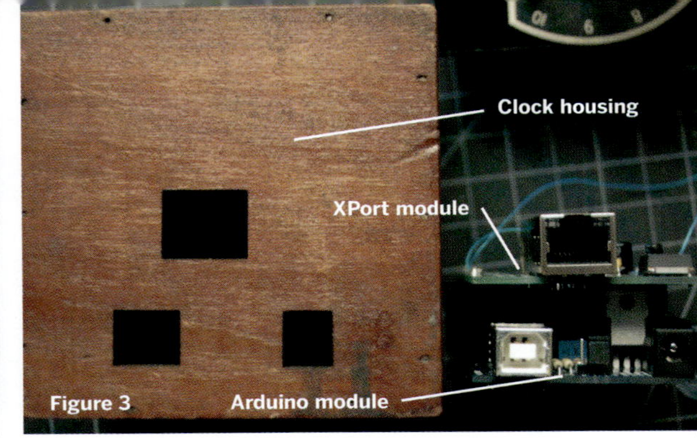

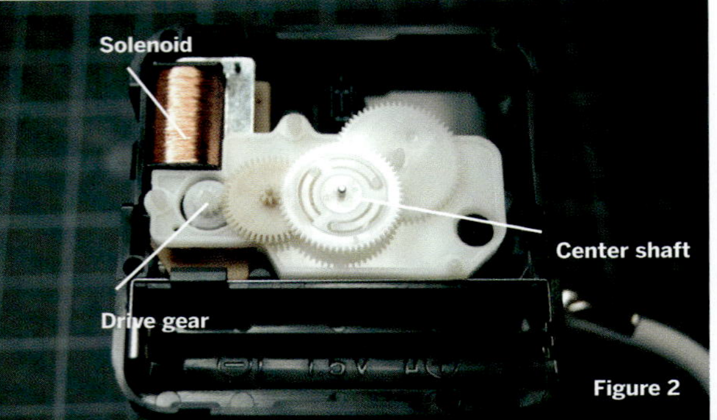

Wires soldered to solenoid contacts. Note crystal at bottom and "mystery chip" under blob in center (Figure 1). Cheap battery clocks all generally work the same way (Figure 2). Rear view of clock, showing holes drilled in housing for Ethernet, power, and the USB-B port for programming the Arduino microcontroller board (Figure 3). Front view of clock, face removed (Figure 4).

MATERIALS

Here's what I chose for the basic building blocks of my email clock.

Microcontroller I used the Arduino module (arduino.berlios.de), a small I/O board and development environment that's built around Atmel's ATMega8 microcontroller. Arduino is based on Wiring, another open and easy development environment, which uses the ATMega128 chip. I opted for Arduino because I knew I wanted to eventually make my own custom circuit board, and the ATMega8 is easier to solder than the ATMega128.

Serial-to-Ethernet Converter Instead of bothering to write my own TCP/IP stack for the Arduino board, I used some handy hardware: Lantronix's XPort serial-to-Ethernet module. These modules can route data between an Ethernet connection and a serial port that feeds a microcontroller. To avoid some soldering, I also recommend the Cobox Micro, which has the same programming interface as the XPort (you simply telnet in) but a simpler physical interface.

Clock I would have loved to build a mechanical clock but didn't have time, so I took the guts out of a cheap battery clock and connected its drive shaft into the hub of a clock-like antique test gauge.

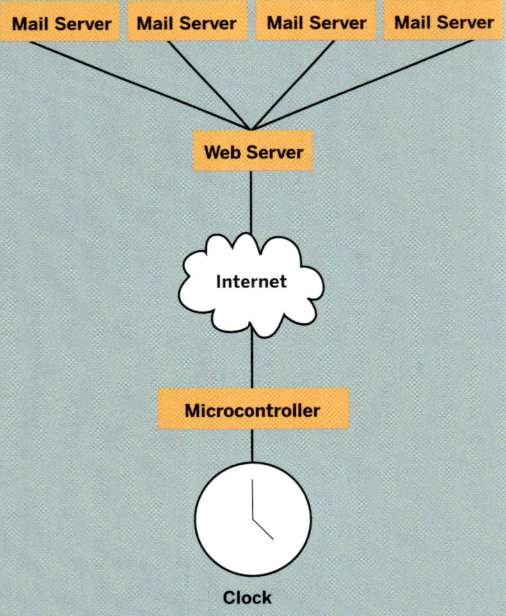

System diagram. A CGI script queries my mail accounts and returns new mail volume. A microcontroller calls the script, and moves the clock's hands forward accordingly.

mystery chip in the middle, sealed in plastic, but I found that the main gear of the clock, which drives all the other gears, is controlled by a solenoid. The two solenoid terminals were pretty easy to spot, and I figured that all I'd have to do was send a pulse through these connections, and the clock would tick happily away.

I soldered a couple of leads onto the solenoid contacts and put the clock back together. Then I pulsed the solenoid directly with 5VDC power, bypassing the mystery chip. The motor jumped, but the clock didn't tick. Clearly, there was more to learn, so I put the clock's battery back in and connected the leads to an oscilloscope to see how the voltage changed as it ticked. The pattern was more complex than I expected. Each second, the pulse would alternate: high-low-zero one second, and low-high-zero the next.

I programmed the microcontroller to duplicate this pattern, and fed its output into the clock. This took some experimentation, but eventually I got the microcontroller to control the clock pretty well. Each tick it generated moved the clock forward about two seconds. Since I wasn't concerned with keeping actual time, this was fine.

The next step was to get the microcontroller to check mail. The XPort needed only three wires connected from the microcontroller side: serial receive, serial transmit, and a reset connection to allow the microprocessor to restart the port. Making a circuit board for the XPort was a challenge because its pins don't follow a 1⁄10" grid, the hobbyist perfboard standard. So I used CadSoft's Eagle software to design a custom board to mount the module into. You can find the layout file at tigoe.net/emailclock.

Then I configured the XPort from my laptop through a USB-to-serial converter cable. Following Lantronix' instructions, I gave it an IP address, gateway address, and subnet mask. I also configured the serial port settings.

I opened a terminal window, telnetted into the XPort, and entered "Hello World!" to confirm that messages were passing through. Then I quit telnet and tried connecting to my web server by entering its numerical address (port 80) in the serial window: `C82.165.199.37/80`.

The XPort confirmed by returning a "C". I responded with an HTTP request for a web page on my server, *http://tigoe.net/pcomp/index.shtml*:

```
GET /pcomp/index.shtml HTTP/1.1
HOST: tigoe.net
```

The server returned the HTTP header and contents of the requested page:

```
HTTP/1.1 200 OK
Date: Tue, 13 Dec 2005 20:50:27 GMT
Server: Apache/1.3.33 (Unix)
Transfer-Encoding: chunked
Content-Type: text/html
<html>
  <head>
```

... and so forth.

Success! Seeing this exchange of HTTP (HyperText Transport Protocol), the normally hidden language of web browsers and servers, meant that I had gotten the XPort to perform as a browser. Now I just had to get the microcontroller to do the same: open a net connection, request a page, and read the results. You can find my code to do this at tigoe.net/emailclock. I uploaded the compiled firmware from my laptop via the Arduino's onboard serial-to-USB converter.

The microcontroller isn't requesting an HTML page — it's calling a common gateway interface (CGI) script, also available at the link above, that checks all of my mail accounts, queries for the volume of new messages, adds the numbers up, and sends the total back to whoever's asking. Since this script isn't expecting to be called from a browser, it doesn't format its results as HTML. To minimize the programming needed on the microcontroller side, it keeps things simple, returning just the HTTP header and one line of text: `<KB: 1234>`.

Once the whole system was working, I found a housing for the clock: a nice, antique piece of electronic test equipment with a hole behind its clock-like face that could accommodate a driveshaft. The clock shaft screwed in easily, and the Arduino module and the XPort fit snugly inside. I drilled a couple of holes in the back for the power and Ethernet cables, and the clock was done.

How well does it work? I still check my email compulsively, but for the couple of hours it took to build this clock, I didn't open my mail program at all!

Tom Igoe heads the physical computing area at the Interactive Telecommunications Program at New York University.

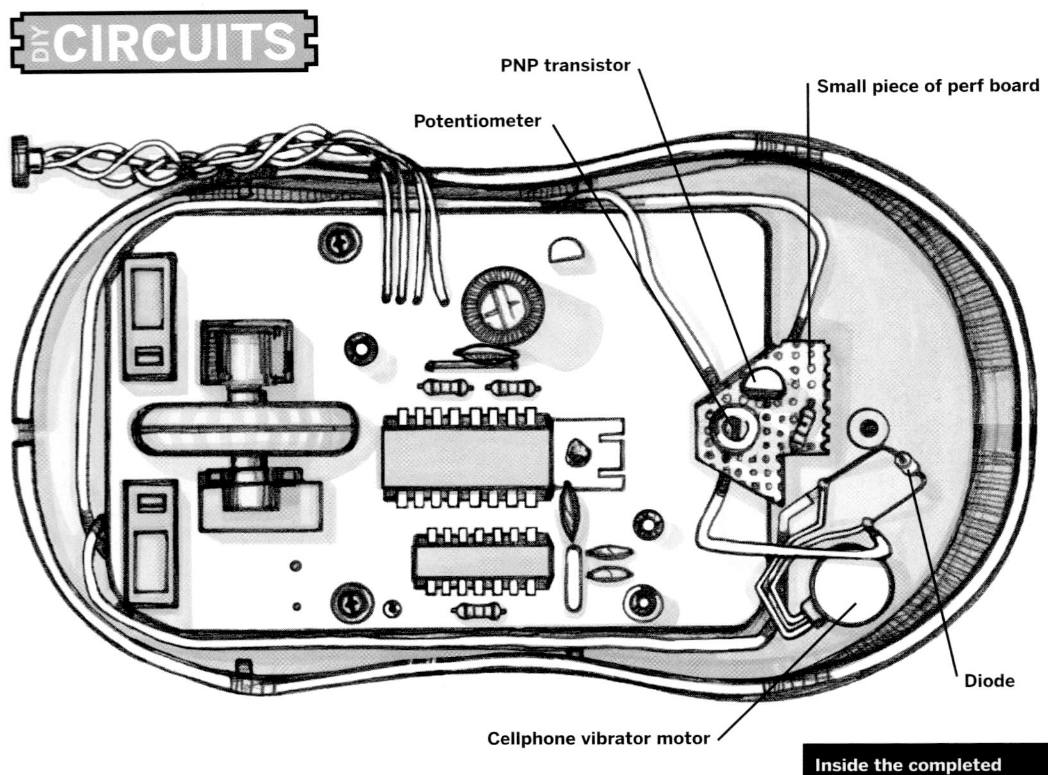

Inside the completed rumble mouse. Note the diode that is soldered in parallel with the motor.

RUMBLE MOUSE

For FPS gaming, a cellphone vibrator gives kick to your clicks.

By Greg Lipscomb

Have you ever been playing your favorite first-person shooter with someone's rumble controller and thought, "I would love to have that capability while playing on my computer"? A friend of mine had a spare rumble pack lying around, so he decided to stick it in a mouse. This inspired me to create my own version.

I determined that my rumble mouse should meet certain specifications. I wanted it to be fully enclosed, with no parts sticking out of the case. It should be an optical mouse, rather than a roller-ball one, connected and powered by USB. In play, the mouse would give a satisfying rumble-recoil when you click the left button — the trigger in most FPS (first-person shooter) games.

I found a cheap suitable mouse from a local surplus store. It had a scroll wheel that I liked, and was large enough to fit the extra components inside. For the rumble motor, I wanted something small, and my fiancée suggested that I use a cellphone vibrator. I had several old cellphones lying around, so I cut one open and located its small motor near the top left corner, which looks like a watch battery with two wires coming out of it.

Naturally, different phones are different. Nokias I've cracked open use small cylindrical motors, not as flat, and in an old flip phone I took apart, the motor was in the same piece as the speaker, close to the LCD screen.

Illustration by Damien Scogin

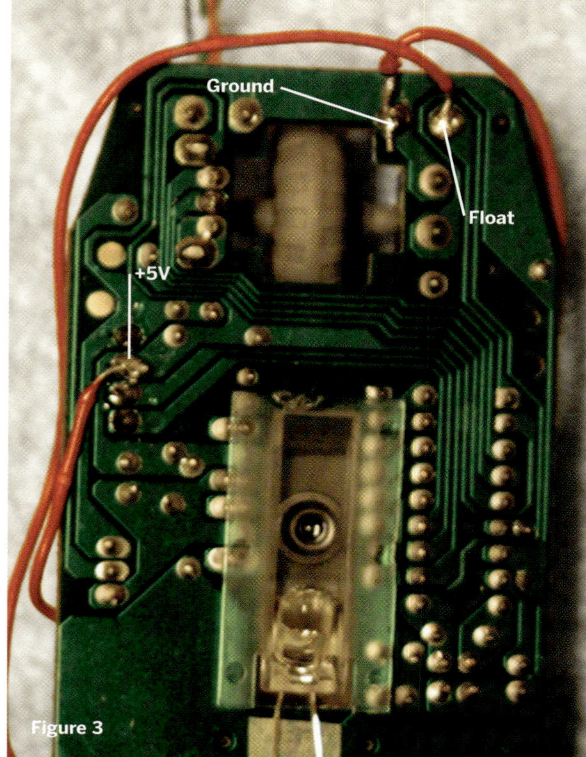

For the rumble motor, I opened up an old cellphone and dug out the vibrator, which looks like a thick watch battery (Figure 1). Bench test for the rumble mouse circuit on a solderless breadboard (Figure 2).

The three pins used in the circuit: +5V is direct from the USB cable, Ground is common ground for the circuit, and Float is the pin that momentarily grounds when you click the left mouse button (Figure 3).

MATERIALS

Optical USB mouse I used an off-brand, model HTM-67WT
PNP transistor, 2N3906 RadioShack #276-1604
Electric vibrating motor Salvage from cellphone, or use Sanko Electric #1E120, available at allelectronics.com, catalog #VB-1
5kΩ or 1kΩ micro trim potentiometer A 10kΩ pot, like RadioShack #271-282, works, but lower values fit the adjusting range better
Diode, 1N4001
1kΩ resistor
Mini perfboard RadioShack #276-0148

I also needed a switch to connect the USB power to the motor when the left mouse button clicks. I chose a PNP transistor for this. The final mandatory part was a 10kΩ potentiometer, which would be used to regulate the current going to the motor.

Find and Adjust the Power

The rumble circuit takes power from the USB port and directs it to the motor, so I needed to find +5V power and ground contacts both on the USB cable and inside the mouse. For the cable, I found a USB pin-out diagram of a USB cable online, and learned that the two outer pins of the USB cable are the +5V and the ground. Then I opened up the mouse and used a digital multimeter to probe for terminals on the board that connected to the USB's power and ground pins. On my mouse, I found that the green wire from the USB was ground, and the blue wire was the +5V. Then I connected the cellphone vibrator directly to the USB power, just to make sure that it could pull enough power to run. It worked perfectly.

I soldered jumper wires to the +5V and ground contacts on the phone, and started putting my circuit together on a solderless breadboard. The vibrator motor had a labeled rating of 3.6V, so I needed a resistor in series with the motor to lower the incoming voltage from 5V. I chose to use a 10k potentiometer, which would let me adjust the voltage and therefore the speed of the motor. You can use your multimeter to make sure that the voltage across the motor is less than 3.6V.

An alternative to the potentiometer is to use two diodes in series. Each diode drops the voltage 0.7V, so if you have two, you would drop 1.4V to give 3.6V over to the motor. Either way, it's important to have a separate diode in parallel with the motor. When the transistor is first activated, the

Photography and illustration by Greg Lipscomb

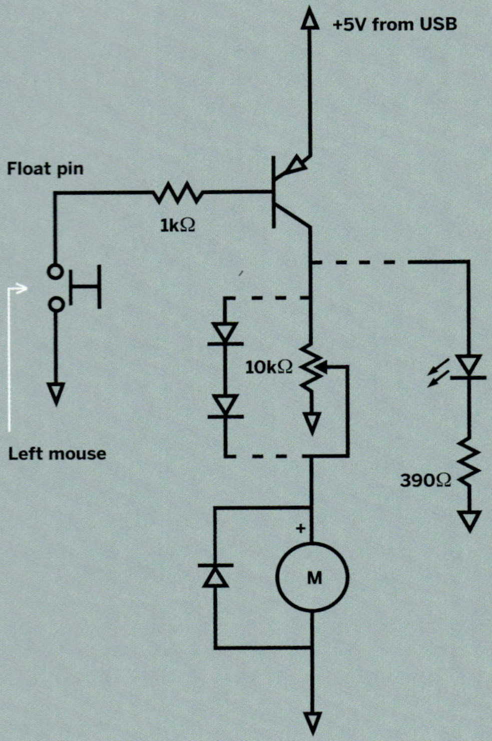

Rumble mouse schematic (optional circuitry shown by dotted lines). Note that the potentiometer can be replaced with two diodes in series, as shown. There is also an optional LED-resistor subcircuit that lights when the left mouse button is clicked.

motor voltage can spike really high, and a diode will protect the transistor from this high-voltage spike.

Intercept the Left-Click

The next step was to locate the board contacts for the mouse's left-click button. It was a single-pole momentary switch, and using my multimeter, I found which pin was ground and which pin was floating. I soldered a jumper wire to the floating pin contact, and continued breadboarding. Since the mouse button operates by grounding the floating pin, I needed an electrical switch that would be activated by a low voltage signal. A PNP transistor was perfect for this requirement. Connect the float pin contact to the base of the transistor, and low voltage from a click action completes the circuit between the emitter (connected to power) and the collector (connected to the motor). Note that this is opposite from an NPN transistor, which is open with a high base current. (Another approach would be to forget the transistor and just hook the motor up between +5V and the floating pin. Then, while the floating pin is grounded, 5V runs through the motor to ground.)

To limit the current flowing through the transistor base, I put a 1k resistor between the floating pin and the transistor base. Then I connected the emitter of the transistor to my +5V from the USB port, and connected the collector of my transistor to the motor through the 10k potentiometer. The other pole of the potentiometer was connected to ground. (When wiring a potentiometer, it is important to attach one wire to the middle pin, and place the other wire on either of the two outer pins. If you connect to both outer pins, you will always get the total 10kΩ.) This completed my circuit on the solderless test board, so I tested it and fixed the bugs.

Put It Together

The final step of my project was to solder everything together. I soldered my components to a small piece of perfboard, using small wires to connect the appropriate leads. Then I used double-sided tape to connect the motor to the casing at the back of the mouse. I also routed the wires to fit nicely into the body of the mouse.

When I completed the assembly, I had a fully functioning rumble mouse that would vibrate on every left mouse click. It is perfect for any first-person shooter game.

Some ideas for improvement would be to add a simple toggle switch between the +5V from the USB and the emitter of the transistor so that the rumble part can be turned off. All in all, this is a straightforward DIY project that is sure to enhance your gaming experience.

Greg Lipscomb is an electrical engineer (Auburn University) who is in his second year of medical school at the University of South Alabama in Mobile. You can see his work at diylive.net.

Since Version 4.9, iTunes has treated podcasts just like content from the big boys.

VIDEO PODCASTING

Producing TV shows on the cheap. By Howard Wen

Video podcasting is easy; it's just like audio podcasting, but you point your RSS feed to video files instead of audio. (*To learn about RSS feeds, see "Podcasting 101" in MAKE Volume 02, page 86.*) For example, in the `<enclosure>` tag, you point to a QuickTime MOV file and list `type="video/quicktime"` instead of pointing to an MP3 and listing it as `type="audio/mpeg"`. And that's it; that's why it's called Real Simple Syndication.

The hard-but-fun part is making the video itself. Here are my suggested tools and tips for making video for podcasts on a budget.

For shooting, I recommend MiniDV cameras, which have fast and easy FireWire download interfaces. A basic model costs less than $400.

Mac software: All you need is iMovie HD, part of Apple's iLife suite (apple.com/ilife, $79). It transfers video from the camera, lets you do simple editing, and exports as QuickTime.

Windows software: Download with WinDV (windv.mourek.cz, free), edit as AVI in VirtualDub (virtualdub.org, free), and convert to MPEG-4 with PSP Video 9 (pspvideo9.com, free). Or download, edit, and save as QuickTime using the Windows version of QuickTime Pro (quicktime.com, $30).

Apps like Final Cut Pro, Premiere Pro, Vegas, and Liquid Edition also do all of the above.

To minimize file sizes, try a few compression settings and compare the results. The most important ones are bit rate, sound quality, and image resolution. 320x240 or 240x180 are best, so shoot everything simple, clear, and high-contrast.

Save in QuickTime format if you can. In my experience, QuickTime files are smaller and better-looking than MPEG-4 and MPEG-1, the other two video formats that iTunes supports.

Whether or not you're already a podcaster, you can make your TV-producer dreams come true.

The typical Linux desktop today looks a lot like a Windows desktop.

DUAL BOOTING LINUX AND WINDOWS XP

Turn your PC into a switch-hitter the easy way, with a LiveCD. By Mark R. Brown

That voice in your head knows that you won't get real geek satisfaction unless you're running Linux, but your practical side knows that you need your favorite Windows applications (read: games). So what can you do? You dual boot. You install both Linux *and* Windows XP.

Are your palms sweating? You've probably heard horror stories about hard drive repartitioning, data loss, and configuration nightmares, but if you take a few precautions, it can be easy.

NOTE: This tutorial covers dual booting with a single hard disk. Setting up separate disks for XP and Linux requires a more convoluted procedure.

Step 1: Pick a Linux Distribution

You know there are at least a million Linux distributions, right? Red Hat. Mandrake. Debian. The names all run together, and each one comes on seven or eight CDs, and takes days to install.

Not necessarily. You don't need a full Linux distribution yet. You want a LiveCD, a distro that runs from a single CD that you can boot from and try out before committing to a full installation.

There's a list of LiveCD distros at frozentech.com/content/livecd.php, sortable by platform and function. You want one that supports x86 hardware, and is listed as Desktop and OS Replacement.

Which should you try? You can start by tracking the herd and looking at the ones with the most votes. Then visit websites, read specs, and look for the features you want. Pay attention to the hardware support. Many distros I researched didn't support my motherboard's sound and video. I eventually chose PCLinuxOS (pclinuxos.com) because it supported both.

Once you've found a few promising distros, download the *.iso* file for each and burn it onto its own CD. You need software to do this, since XP won't write *.iso* images directly. I used Nero Burning ROM, which came with my CD burner. You can also use ISO Recorder Power Toy (isorecorder.alexfeinman.com, free). Boot from each CD, and test them out. Find one that works with your hardware, internet connection, and peripherals.

You can stop here, and just boot from CD when you want to run Linux. But what kind of a geek does that? Time to repartition your hard drive.

Step 2: Partition Your Hard Drive

Here's the problem: Your Windows XP is installed on a partition formatted in NTFS. Most Linuxes can read NTFS but can't write to it. Meanwhile, XP can't read a Linux partition at all. So any data you want to be readable and writable from both XP and Linux needs to be moved to a separate partition formatted in FAT32, which all OSes support.

And yes, you're going to have to reinstall XP, so back up all of your data to CDs. Burn it *all* — there are hidden address files, configuration files, cookie files, and who knows what else on your hard drive that you will regret not having later.

Boot from your Linux CD and find the partitioning program. Mine was called `qtparted`, but your mileage may vary. Use it to subdivide your hard disk into four partitions, like this:

TYPE	FILE SYSTEM
Windows	NTFS
Linux	Linux
Linux swap	Linux swap
Data	FAT32

Make sure they're in this order. How big you make them depends on several factors. Don't give XP any more or less than you really need for XP itself (1.5GB minimum) plus all of your Windows applications. I generously allocated 7GB because I use some big programs like Visio and Paint Shop Pro.

Your Linux partition should start no further up on the drive than 1,024 cylinders (about 7.8GB). Read your LiveCD's help files to find out how much disk space your Linux distro needs, and how much to allocate for swap space. Mine needed 10GB for the Linux and 3GB of swap space. Allocate all remaining space to your data.

Step 3: Install Windows XP

Remove your Linux LiveCD and reboot with your Windows XP installation CD. Install XP. When it asks if you want to make a recovery diskette, say "Yes" and make one. Be safe.

After installation, remove your Windows CD, boot to XP, and make sure that it still works OK.

Step 4: Install Linux

Use your LiveCD to reboot to Linux, and read the help files for instructions on doing a full install.

The installation program will ask you which partitions to install to for Linux and Linux swap. Make sure you point it to the ones you allocated when you repartitioned your drive. Note that whereas Windows refers to the first hard drive as `C:`, Linux calls the same drive `hda`.

The installer will probably also ask you to configure the boot loader program (usually GRUB or LILO), which allows you to choose your OS on startup. My install ran LILO and auto-detected both my XP and Linux partitions, but you may not be so lucky; you'll find out soon. Install your boot loader in the MBR, or Master Boot Record, so it runs automatically every time you reboot.

Remove your Live CD and reboot. You should see a menu letting you choose XP or Linux. But if your computer boots to XP without seeing Linux, you'll have to find and run your boot loader and manually point it to the Linux and XP partitions.

Step 5: Restore Your Data

Finally, to restore your data from the backup CDs, just drag and drop from either Windows or Linux.

Of course, you'll also need to reinstall your Windows apps, configure, customize, and get them to see your hardware, and also learn Linux. But that's what being a computer geek is all about, right?

Once upon a time, Mark R. Brown was managing editor of *.info*, the legendary Commodore computer mag. He also wrote lots of now-out-of-print books about the internet for Que.

DA XEROX CODE

How to read the tracking dots in your color Xerox printer. By Annalee Newitz

Early in 2005, Internet Archive founder Brewster Kahle noticed something a little weird about the color copies he was making. When he looked at the pages really closely, he could see hundreds of tiny yellow dots — and they looked like they formed some kind of pattern. He recalled seeing an article in *PC World* about how the Secret Service busted counterfeiters by using marks on the fake bills that could be traced back to the printers that made them. If these dots were a method of tracking, then Kahle wanted to know exactly what information they contained.

He turned to Electronic Frontier Foundation technologist Seth Schoen for answers. Schoen, with the help of EFF intern Robert Lee, combed through hundreds of pale-yellow-speckled color copies from Kahle's machine. They contacted a Xerox research fellow, who confirmed their worst suspicions: yes, the dots were for tracking. How widespread was this practice? Was it just Xerox?

To find out, Schoen posted a call for color copies on the EFF website. He soon received hundreds of printouts from all over the world. "One was from an ice cream parlor in Australia," he recalls. "Another was from a symphony orchestra." All of them had the telltale yellow dots.

It turned out that companies like Dell, IBM, Canon, Lexmark, Toshiba, and many others were creating the dots on color laser printouts (for a list of printers that generate tracking dots, see eff.org/Privacy/printers/list.php). Schoen speculated that the nearly

Barely visible dot-matrix "fingerprints" enable any page to be traced back to its color printer.

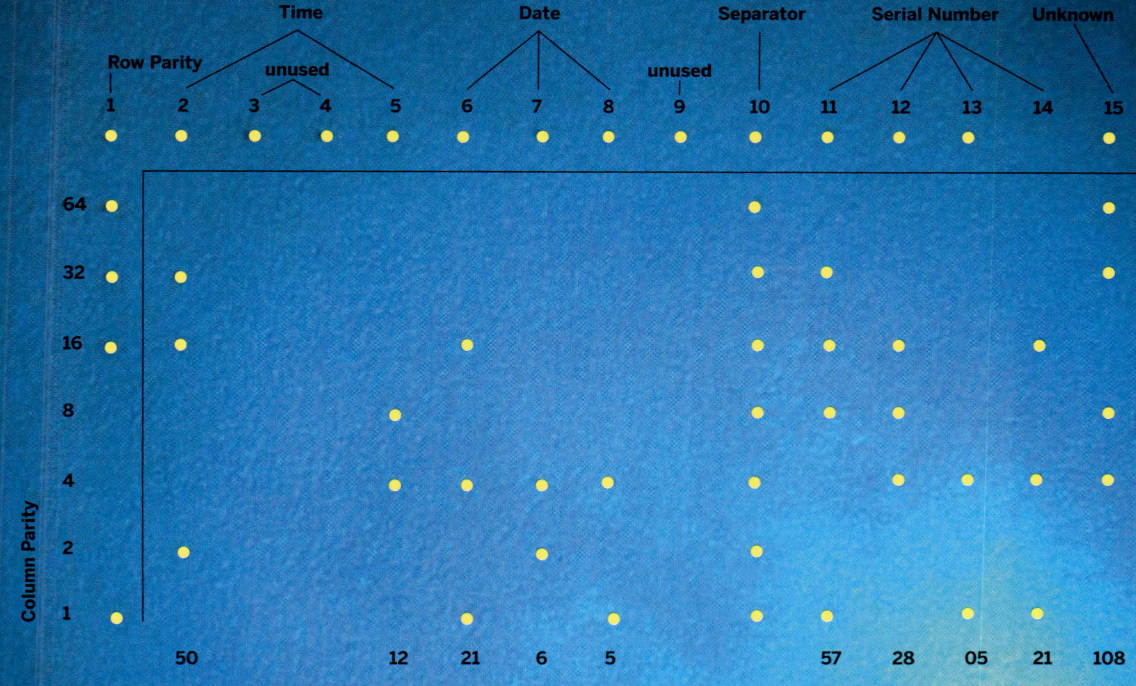

This repeating dot pattern encodes the DocuColor printer serial number, the date and time the page was printed, and other bits of info not publicly known. Parity bits along top and left keep the rows and columns odd (except top row). Rows carry binary digits, numbered up from the bottom. Columns 6-8 encode date (DMY order), 2 and 5 encode time (minute, hour), and 11-14 encode product serial number, two decimal digits per column.

Image by Annalee Newitz and Seth Schoen

invisible patterns — which seemed to repeat several times on a given page — must encode the serial number of the machine that prints them, possibly along with other information such as a time/date stamp.

Blue Light Reveals

By now, Schoen's office at EFF was awash in printouts, and he'd become adept at finding patterns in the dots by shining a blue LED on each page while it was magnified ten times under a USB microscope. The blue light made the yellow dots look almost black. When magnified, it became clear that almost every printer uses a repeating dot pattern that forms a grid of 15 columns and 8 rows. It also seemed obvious that the dots probably represented binary numbers. But which direction should the numbers be read? And how were digits being represented? Was each column one number, or part of a larger number? Schoen was stumped.

Luckily, a cryptanalyst came calling. Swiss graduate student and crypto expert Joel Alwen took one look at the dot patterns he found on Schoen's desk, and knew he had to solve the puzzle. He decided to work on dots produced by the Xerox DocuColor 12 because Andrew "Bunnie" Huang had posted some pictures of them in his blog.

"It took me about five or six hours total to decode it," Alwen says. After looking at several pages made by the same printer at different times, Alwen made his breakthrough. "The watermarks only differed in column 2, so I guessed that this was a timestamp that was arranged with low order left and high order right."

Once he knew the numbers read from right to left, he quickly discerned that the next few columns were the date, and the columns after that spelled out the serial number (*see "Read the Dots," next page*). He and Schoen also realized that the first column and first row were essentially parity bits — they speculate that this may be the case for dot grids created by most printers.

Right now, only the Xerox DocuColor 12 dot code has been cracked completely. Dell dot patterns are similar to Xerox, so Schoen speculates it will be the next code broken. That leaves dozens of other dot patterns from other printer manufacturers unsolved.

How to Read the Xerox Codes

With a few of the tools and techniques developed by Schoen, Huang, and Alwen, you'll have everything you need to find the dots, analyze them, and (hopefully) crack a few more of the encoded patterns. Before you delve in, though, you might want to play around with reading the Xerox codes.

To read the Xerox codes, you'll need a few items.
Color printout from a Xerox DocuColor 12 printer The dots look best on the white portions of the page, so make sure your printout has lots of white space.
Scanner or USB microscope Schoen recommends the $60 Digital Blue for Windows and a blue LED.

One method for finding the dots, of course, is simply to shine the blue LED on the paper and try to see them with your naked eyes. Schoen is able to see them this way, but I find it impossible. Most people use magnification or image enhancement.

Using a Scanner

If you want to go the scanner route, simply scan a chunk of the page at a slight magnification, and dump it into Gimp or Photoshop. Fiddling with the color balance and background noise will allow you to turn the yellow dots gray or black. You can see some excellent enhanced scanned images of the dots at makezine.com/go/dots1 and makezine.com/go/dots2.

If you want to add on another project, try building Huang's blue light scanner, which makes the dots look extremely clear, bunniestudios.com/wordpress/?page_id=51.

Using the USB Microscope

You don't need to see the whole page, since the dot pattern will repeat several times. A good bet is to magnify it to 10X. Once you've got the paper positioned in the microscope, shine the blue LED on it. You should see a nice example of the dot pattern fairly clearly, with the yellow dots turned a dark gray.

Read the Dots

Remember that the Xerox pattern has 15 columns and 8 rows. To find the grid, look for the only row with an even number of dots; that's the topmost row, the horizontal parity. All the other rows (and columns) are odd. Once you've located the pattern on your sheet, save it or write it out on graph paper. If you know how to read binary, you can use the EFF's guide to decode each column (read right to left):

15: Unknown (often zero; constant for each individual printer; may convey some non-user-visible fact about the printer's model or configuration)
14, 13, 12, 11: Printer serial number in binary-coded decimal, two digits per byte (constant for each individual printer)
10: Separator (typically all ones; does not appear to code information)
9: Unused
8: Year that page was printed (without century; 2005 is coded as 5)
7: Month that page was printed
6: Day that page was printed
5: Hour that page was printed (may be UTC time zone, or may be set inaccurately within printer)
4, 3: Unused
2: Minute that page was printed
1: Row parity bit

Alternatively, you can use a handy program that Schoen wrote that automatically decodes the dots for you. It's at the very bottom of the page at eff.org/Privacy/printers/docucolor.

Now you should be able to figure out the serial number of the printer that created your page, as well as the date and time it was made. (Note: Some machines print the wrong date/time because they've been set up incorrectly). The next step, for inveterate geeks and makers alike, is to start cracking codes on other printers. You can use the same procedure for seeing the dots, but keep in mind that some dot patterns could be bigger or oriented differently than the Xerox pattern. Some are probably read bottom to top, or left to right. Binary representation may vary.

The point of this project — aside from hours of amusement — is to defang a tracking system that can be quite privacy-invasive. What if you want to print political flyers anonymously? Or you don't want people to know when you printed a particular document? The more we know about these dot patterns, the better we can protect the anonymity and privacy of innocent people.

So get to decoding!

Annalee Newitz (techsploitation.com) writes a syndicated column called Techsploitation, is a contributing editor at *Wired* magazine, and is a huge fan of Upper Paleolithic technology.

Use a couple of white LEDs to illuminate the original Game Boy Advance's display.

LET THERE BE (FRONT) LIGHT

Add an LED front light to your Game Boy Advance for $5. By Dave Prochnow

One of the biggest sore points with the original Game Boy Advance (GBA) was its lack of a backlit screen. You and your GBA could while away many waking hours, but with the arrival of dusk, the fun faded as quickly as *Donkey Kong* high scores.

MATERIALS

Game Boy Advance

White T-1 LEDs (2) I got mine from goldmine-elec.com, part #G155b2.

10-ohm, ¼-watt resistors (2)

Red and black insulated wire

Nintendo rectified this failing later with the release of the Game Boy Advance SP, which features a lighted LCD.

Barring the purchase of a new GBA SP, what can a lowly GBA owner do? The Afterburner kit from Triton Labs made GBA screens glow with an ingenious, clear, front-lit LED panel, but that kit has been discontinued. Here's a simpler and rougher solution, inspired by Benjamin Heckendorn's book *Hacking Video Game Consoles* — just hack a couple of white light-emitting diodes (LEDs) that draw from the GBA's own power and illuminate the screen from inside its case.

Finding the parts and tucking them inside the

Photography by Dave Prochnow

Underneath this protective mat is the LCD. The two white LEDs will be placed in opposite corners along the bottom (Figure 1). Wiring harness and LEDs in position on GBA (Figure 2). LED cathodes connected by black wire, anodes connected to red wire through resistors (Figure 3). The toughest solders are hooking up to power and ground, S2 and GND, on the game cartridge interface board (Figure 4).

GBA is easy; the hard part is getting the case open in the first place. In what must have been an attempt to tick off hackers, Nintendo assembled the GBA with six tamperproof Triwing screws. You really need to use a Triwing screwdriver to remove these; you can't simply force 'em out with a regular screwdriver. One option is to get the Craftsman 32-piece Security Screwdriver and Bit Set from Sears (item #00947486000, $25), which includes four Triwing bits, plus Torx and others.

If buying a new, special-purpose screwdriver to work on a vintage game console seems ridiculous, take your GBA into your local electronics repair shop. If you're lucky, they'll have some Triwings, and you can replace the tamperproof screws with regular Phillips screws later.

Once inside the GBA, it's a simple two-screw (Phillips screws) process to separate the main circuit board from the display area. Carefully peel away the protective gray cushion from the lower bottom edge to gain access to the LCD itself. Then you should be able to pry the LCD up and out with a gentle push from a thin screwdriver.

The LCD is supported and protected by a black rubber frame covered with an adhesive film. To make room for the white LEDs, you need to remove the lower right and lower left corners of this frame. Do this carefully! I used a sharp knife to gently cut two rectangular pieces out of the rubber.

Place the diodes in the corners with their legs sticking out. Then assemble the wiring harness by soldering one end of a small length of red wire to both of the 10-ohm resistors, forming a T.

Connect the negative/cathode (shorter) legs of the two LEDs with a piece of black wire, and solder another piece of black wire to either one of the cathodes. Solder the LEDs' positive/anode (longer) legs to the free ends of each resistor.

Now, the tough part: carefully solder the other end of the black wire to the pad labeled GND on the GBA cartridge interface board, and solder the red wire to the pad labeled S2.

Finally, realign the main circuit board, lower the speaker back into its receptacle, tidy up your wiring, make sure all the buttons line up, and screw everything back together. You now have a new, bright-eyed GBA that will still be able to play all of your fave games. Game on!

Dave Prochnow wrote "Panzeroids" on page 58.

Young kids have more fun playing CD-ROMs using a touchscreen.

STICKY FINGERS

Modifying tech for pre-mouse toddlers.

By Damien Stolarz

As anyone raising a toddler knows, baby minds are hungry sponges for information. As a parent, I try never to pass up a learning opportunity. I'm a big technology enthusiast, and so naturally I want my kids to know everything about the hardware and software that runs our world.

Many parents are (understandably) overprotective of their devices at the expense of their children's education. They worry that kids will put sandwiches in the VCR or even shock themselves. What they fail to realize is that only through working with objects and technology — by making mistakes and learning from them — will their children ever gain the necessary knowledge and competence. And in my own case, as a maker and tinkerer who has broken more than my fair share of gadgets in the process of learning how they tick, it would be downright hypocritical of me to stop my children from breaking all my digital devices for the noble cause of their education.

Toddler Kiosk

As our daughter grew older, she wanted to imitate what her parents did. She liked to sit in front of our PCs, bang on the keyboard, and inform us that she was "working."

We bought her a brightly colored PS2 keyboard that has a picture next to every letter (apple for A, snake for S, dog for D, and so on). For a while, we would just pull out her keyboard and let her type on it when she wanted to work along with us.

After a while, we knew she needed her own computer — heck, she was already 2. So I dusted off an old G3 Macintosh and bought a handful of

Photograph by Damien Stolarz

games designed for toddlers.

However, I saw how much she struggled with the so-called "intuitive" mouse user interface. She would move the mouse to the screen with no results; she would look at the mouse and move it, then look at the screen, lifting the mouse. With Mom and Dad's help, she got very interested in several of the interactive games, but we always had to be there and work the mouse for her.

I was familiar with a series of VGA touchscreen monitors, from a company called Newision.com, that were used for in-car computers. These 7" and 8" monitors are designed to stick to the dashboard of a car, but I found that they worked just as well stuck to the top of this G3 Macintosh. Since they provided a Macintosh touchscreen driver, I simply hooked up the display, and my daughter was immediately able to play the games all by herself.

I rigged up a launcher program that had big buttons, so she could even launch the games herself. And since one of the games asked her to type in her name (and I had the foresight as a parent to give her a name with only three letters), she quickly learned how to hunt and peck her name in and press Return.

Mobile Entertainment

One of my hobbies/professions is car hacking. Instead of purchasing the $1,800 factory DVD option with our new Dodge Caravan, I installed a Mac mini ($599) and a fold-down 15" touchscreen monitor ($199), so that my daughter could watch *Dora the Explorer* in hi-def resolution.

The essential feature of in-car entertainment systems these days is "multi-zone" entertainment, meaning the kids in the back have headphones so the driver and front passenger can listen to their own audio program.

I initially tried giving her headphones, but you will find that babies take a while to warm up to hats, headphones, sunglasses, or anything else you try to wrap around their heads. I wanted to find a way to create an isolated sound environment for her so we could listen to our own music.

Most toddler car seats consist of a washable, foam-filled padded cover wrapped around the plastic frame. Inspired by a noise-canceling headphones project (*MAKE, Volume 05, page 56*), I took the connecting loop out of a pair of inexpensive headphones, duct-taped them to the left and right sides of the plastic car seat, and then put the foam back on. For $5, I had upgraded my daughter's car seat with a sophisticated "surround sound for babies."

I used a mini-jack extension cord to connect this to an in-car DVD player. Now, I can have my own Mac mini-based music in the front seat, and my daughter can watch her DVDs and hear them in the back seat. Because of the perfect positioning of the speakers on either side of her head, she has her own aural experience, apparently unaffected by the music in the front seat.

Toddler Shuffle

A toddler's rabid affinity for media makes her able to listen to the same song repeatedly, apparently indefinitely. On one of our daughter's CDs, she has found a two-minute song that she likes to hear over and over.

Unfortunately, the CD player in our car does not have a "repeat song" function, and despite the integrated steering wheel controls in the minivan, it is still unnerving to have to press "back" every two minutes during long drives.

I thought about getting my daughter an old iPod, but the interface would be too complicated. Fortunately, Apple came out with the iPod shuffle, which has no breakable screen, is small, and more importantly, has a very simple three-button, one-switch interface.

We had also by now acclimated her to headphones — not earbuds, but a small pair of over-the-top earphones.

We showed her that to get her music, she needed to move the switch on the back of the iPod (to turn it on) and then press the "play" triangle. And when her favorite song comes on, she can simply press the "back" arrow to repeat it.

She's still getting the hang of the iPod shuffle — she'll overshoot, rewinding too far — but we don't have to intervene nearly as much now. And when she needs help, we can usually talk her through how to get back to her song again ("Press the left arrow two times. Is that your song? Good."), and she can do it all by herself.

Damien Stolarz is an inventor with a decade of experience making different kinds of computers talk to each other. His book *Car PC Hacks* is published by O'Reilly Media, Inc.

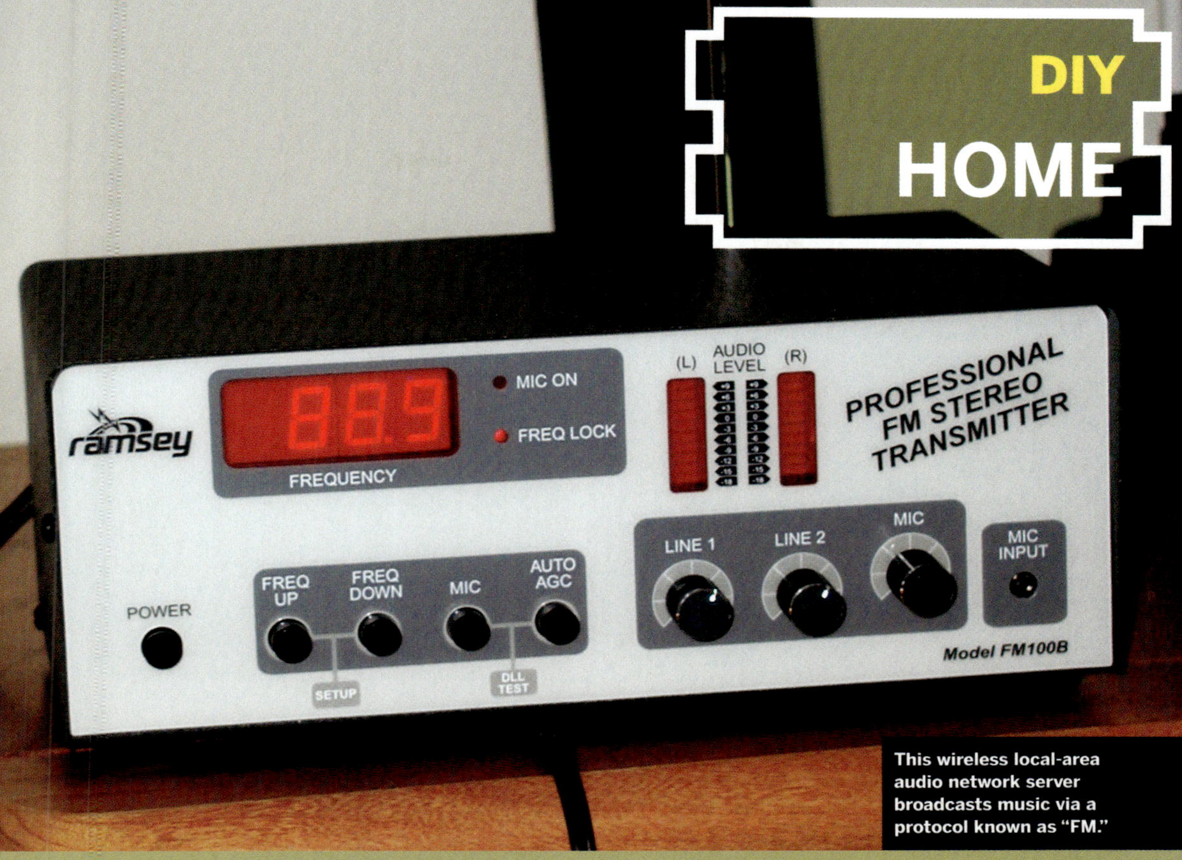

This wireless local-area audio network server broadcasts music via a protocol known as "FM."

HOMECASTING DIGITAL MUSIC

Good old FM beats wi-fi for sending streamed music around your house. By William Gurstelle

Photography by William Gurstelle

I subscribe to Real Networks' internet-based music service Rhapsody, which lets me stream a wide variety of music through to my computer's speakers. Digital music services like Rhapsody access millions of songs at the touch of a computer mouse. At ten bucks a month for unlimited use, this seems like a deal. The problem is, I don't listen to music only at my computer — I want to listen on my living room stereo, in the driveway as I work on my car, in the kitchen, in the basement, and so on. You need to untether the music from the computer's often-mediocre sound system in order to get the full benefit of your customized, on-demand music programming.

Rhapsody recommends using a multi-room digital music system call Sonos (sonos.com), but it can cost thousands of dollars. Products like Sonos distribute the music stream via Ethernet, either cabled or wireless. This makes listening outside problematic, because even if I have wi-fi set up to transmit, I still need a digital device and electrical power out where I'm listening. Using wi-fi also makes it impossible to listen using just portable headphones.

Legal FM Transmitters

A more cost-effective and flexible solution is a small FM radio transmitter. This trick is familiar

The completed power supply assembly from the Ramsey FM100B Super Pro FM Radio Station Kit.

to iPod users who play songs through their car sound systems, but broadcasting throughout your house and yard requires a bit more power.

Legal FM transmitters are a nearly perfect solution. They are relatively inexpensive, and they use the most ubiquitous of music players — FM radios.

I tried a few of these transmitters, starting with the C. Crane FM transmitter ($70, hobbytron.com or amazon.com). The Crane was plug-and-play simple, but its tiny size was matched by tiny power. Its out-of-the-box range was just a few feet. To increase this, I peeled off the metallic FCC label on the bottom, and used a screwdriver to turn the small potentiometer hidden underneath. But even at its max, the Crane's range is limited and will reach only a few adjacent rooms.

I also tried the much larger Ramsey FM100B ($250, hobbytron.com), which puts out a quarter watt of power, the maximum the FCC allows for unlicensed radio transmitters.

Some Assembly Required

The FM100B comes unassembled, as a large kit with over 800 solder points. According to the instruction sheet, it takes ten hours for an advanced hobbyist to assemble the kit, and 24 hours for a beginner. Given the number of components and the amount of soldering, these estimates seemed optimistic.

Unfortunately, the kit I received contained a bad volume control and a bad fuse holder. This added some troubleshooting time to the build, but the company was responsive and replaced the defective parts quickly.

I enlisted the aid of my friends Al and Rick, who had previous experience building Ramsey FM transmitters. With a kit this complex, they knew that we needed to set up an efficient assembly line. They prepared a work area with adequate lighting and work surfaces, and then began work in earnest, systematically identifying components (resistors, capacitors, switches, integrated circuits, etc.), soldering them into place, checking items off the list, and testing connections.

The kit directions divide the assembly process into discrete sections organized by subsystem, including the display board, mixer, filters, and transmitter. Each section clearly documents what needs to be accomplished, and builders with reasonable soldering skills should have no problem

My friends Al and Rick, who had built FM transmitters before, set up an efficient assembly line; Rick wrangled the components, and then Al soldered them into place (Figure 1). The Ramsey FM100B's main board, in the early stages of assembly (Figure 2). The completed transmitter, ready to fill the airwaves (Figure 3).

following along. As a bonus, the manual provides technical background as to what the subsystems do and how they work, turning assembly into a learning experience.

Many hours later, the kit was assembled and debugged. It was time to try it out. The FM100B has several input options. I hooked my computer into its stereo RCA jacks with a mini-to-RCA adapter cable ($8 at RadioShack).

The next step is to select a broadcast frequency, using the digital frequency selector on the front panel of the transmitter. I live in a big city, where the FM spectrum is crowded, but I found a fairly empty space toward the lower end of the FM band. I dialed it into the FM100B, launched Rhapsody on my PC, queued up several songs, and started broadcasting. Then I tuned the FM radios in my house to the broadcast frequency. The sound on several radios was pure and clean, as clear as any other FM station. A few radios did not work as well at first, but after fiddling with the placement of the FM100B's antenna, I decreased or eliminated the static and fade.

There is definitely a feng shui aspect to antenna placement, as the proximity of walls, windows, metal, and concrete have a big effect on the transmitter's ability to "hit" individual receivers.

The FM100B has a microphone input and mixer knobs, so you can do sophisticated radio station tricks. Assembling and troubleshooting the FM100B was a great learning experience, and it gave the desired result: freedom of movement indoors and out, without interrupting the music.

DON'T PUMP UP THE VOLUME

Home FM transmitters have low power because Federal Communications Commission regulations restrict the ability of any device to cause interference with legal radio signals, especially those put out by licensed stations. The maximum power an unlicensed transmitter can legally develop is 0.25 watts of signal. That's not much to work with, but it keeps the radio spectrum free from interference. Compare this with the 110,000-watt FM signal radiating from a big station such as Los Angeles' KPFK or San Francisco's KQED.

William Gurstelle's latest book, *Adventures from the Technology Underground: Catapults, Pulsejets, Rail Guns, Flamethrowers, Tesla Coils, Air Cannons, and the Garage Warriors Who Love Them*, came out in January.

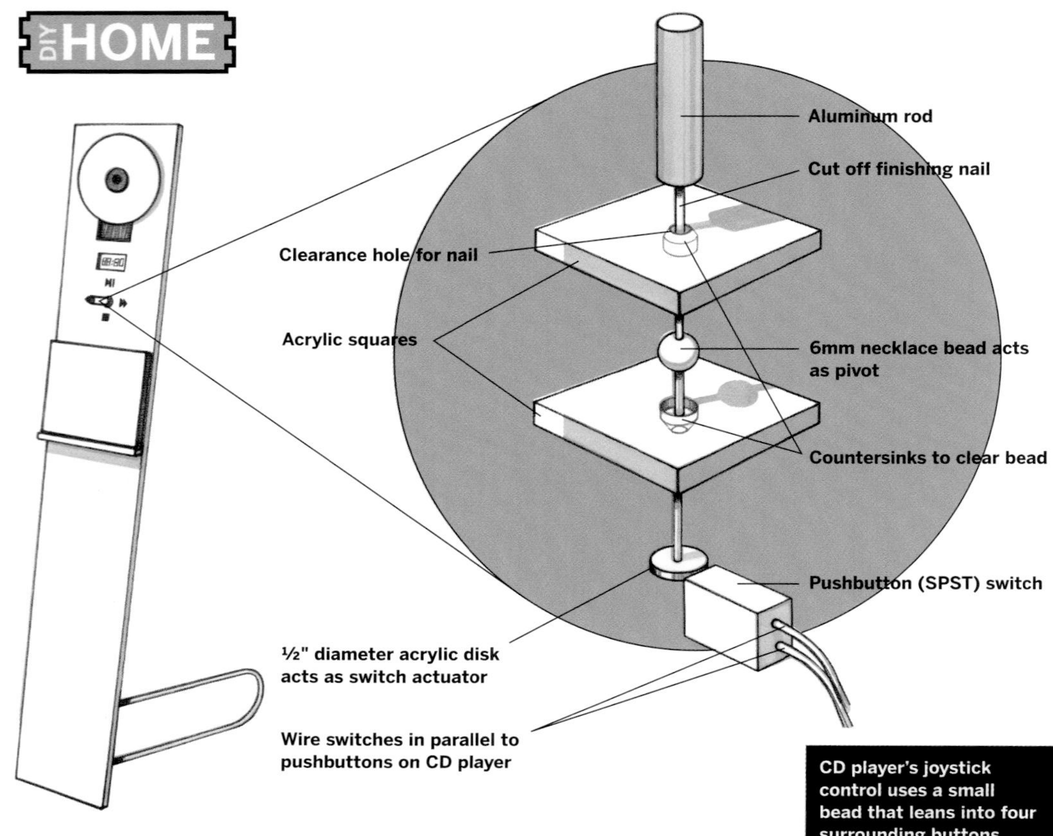

CD player's joystick control uses a small bead that leans into four surrounding buttons.

BORN-AGAIN BOOMBOXES

Up from the ghetto blaster, a new life as a functional-art stereo component. By Larry Cotton

Don't you love the fact that off-brand, bait-and-switch consumer electronics can be so cheap? I've seen portable CD players for under ten bucks, including headphones, and you can get a commodity CD boombox for about $30. Unable to resist such great values, I've bought a few of these beauties to see what makes them tick, and then reconfigured their aesthetics and turned them into attention-getting CD decks.

The procedure involves taking the boombox apart, and deciding which components will live and which you'll save for who-knows-what later. Then you arrange the keeper components onto some structure, which can range from a beautiful piece of hardwood to a coat hanger. Rather than keeping them as portable stereos, I've been turning the boxes into deconstructed CD players to use with headphones or plug into an amplifier.

This means dispensing with the speakers and wasting the FM tuner, but it's still worth starting a project like this with a boombox rather than a cheaper portable CD player, because its modules are larger, more robust, and more forgiving when you rearrange and attach them to strange objects.

I'm especially proud of a reconfigured boombox that I've named *The White CD*, a play on the Beatles album that's commonly (and erroneously) referred to by a similar title.

The White CD's innards are interconnected with relatively long wires, which lent themselves to a vertical arrangement. The CD spindle could be on top, the display would be below that, and — ta da! — a joystick below the display.

A joystick? In previous boombox mods, I kept

Illustration by Damien Scogin

Original cheap boombox on the half-shell, ready to resurrect as *The White CD* (Figure 1). Prepare the mounting board area, dug out for the display (Figure 2). Finish the surface and fill in cracks and seams with automotive body filler and glazing compound (Figure 3). Bend heated Plexiglas base over wooden form and hold until cooled and set (Figure 4).

the original push buttons for operating the box, but with *The White CD*, I wanted to do something different. I made a joystick and wired its switches in parallel with four of the switches on one of the box's circuit boards. I didn't modify any of the other circuit boards or the power transformer.

You should retain the box's headphone jack and make it accessible on the finished piece; that's how you'll plug it into a stereo amplifier, via mini-to-RCA cable. If it's destined to be a component for your stereo and you won't be using headphones, don't bother making the volume knob accessible, since you'll be using the volume on the amp. I've found that turning the CD player's volume control almost all the way up matches the output of a typical component CD deck.

A minimal FM tuner is usually included on the box's amplifier/power circuit board. You can incorporate it if you wish, and if so, be sure to make the CD/FM switch accessible.

The boomboxes I've reconfigured play standard CDs with reasonably good fidelity. For more bucks, you could choose boxes that will also play MP3-encoded discs, CD-Rs, and even audiophile SACDs and DVD audio disks.

Photography by Larry Cotton

Making *The White CD*

Here's how to "transmogrify" a boombox into *The White CD* configuration. As you work, test that the player still functions after each change you make, and then disconnect from power before moving onto the next step.

NOTE: The lasers in CD/DVD players can be dangerous. Don't stare at the laser when the device is on! Also, never rout or drill near digital circuitry, especially on the board that controls the CD stepper motors. Flying sparks can fry sensitive chips.

1. Carefully take your boombox apart without breaking or disconnecting any wires. Be especially gentle with the assembly that carries the turntable, stepper motor, and tracking mechanism. You'll likely find three circuit boards: the control electronics board, the display/switches board, and the amplifier/power board. Watch out for all of them.

I've seen two designs for securing the CD to the spindle. Some players use three spring-loaded plungers at 120-degree angles. Others use a ring magnet in the lid. If yours uses a magnet, be careful when removing it because it could be fragile.

Circuit board and transformer from donor boombox attached to mounting board with standoffs and heat-shrink tubing to prevent shorts.

2. Make a mounting board that's big enough to hold all the pieces you're using. Hardwood looks cool, but *The White CD* is made of primed masonite laminated to plywood. You'll also need auxiliary mounting posts and bosses.

3. Cut a hole from the front of the mounting board, then rout a recess in the back of the board deep enough to let the CD just clear the front surface. This lets you really see and get to the CD, which I like.

4. Mount the turntable assembly to the board using some sort of vibration-damping material in between. You could use the grommets originally attached at the corners of the turntable, or glue it down (or in) with bathroom caulking compound.

5. Mount the display/switch board in a similar manner.

6. Mount the CD control electronics board using special care, noting that certain wires must be able to flex as the CD plays.

7. Mount the other circuit boards and transformer with screws and possibly short pieces of tubing as standoffs, if necessary. Disconnect any unused subsystems, such as speakers. Make sure all 115VAC wires are well-insulated. Keep all circuit boards away from each other so that no short circuits develop.

8. If you need to switch the CD deck on and off independently of the other components in your stereo system, add a simple, single-pole, single-throw (SPST) switch that can handle a few amps, placing it in series with the power line cord and the transformer.

9. Connect the CD player to your sound system, put a CD on the spindle, add the magnet (if used), and ensure clearance for the CD. Power it up.

If there is an FM radio circuit, be sure that the CD/FM switch is in the CD position. Turn up the volume on both the player and the amplifier. By now, you should be hearing deafening sounds from the Stones or possibly John Prine.

10. Now you can build and connect the joystick, wiring its switches in parallel with the boombox's buttons (*see the diagram on page 136*). My joystick functions are: Up = Play/Pause; Down = Stop; Right = Next; and Left = Previous.

On the control PCB, find the leads that the buttons connect to, and solder in the same connections through the joystick. Be prepared to spend a fair amount of time figuring out where the joystick's switches should connect to before soldering them in.

11. Make front-side guides for the joystick, to prevent two diagonal switches from being actuated simultaneously.

12. A joystick makes the boombox's original switches redundant, so I covered them up with a scrap piece of 1⁄16" Formica. Make sure you can still clearly see the CD track number.

13. Make a stabilizing bracket for the mounting board from ¼"-thick x 4"-wide clear acrylic (Plexiglas), by carefully heating the acrylic over an electric stove burner and bending two 120-degree angles. Strategically place aluminum foil to reflect heat as necessary, and control the bend angles with a fixture.

14. Glue a jewel-box shelf to the mounting board about 7½" below the joystick centerline.

15. Unless you go with hardwood, finish your baby with Bondo body filler (available at auto stores), automotive glazing compound, and paint. Fill large flaws with body putty and smaller ones with glazing putty. I then painted the back flat black.

16. Finally, use a hobby knife to carefully cut vinyl tape (available in black and colors) into the graphic shapes that indicate the joystick functions. Precisely locate the shapes around the joystick, and press into place. You're done!

17. OK, I lied — you're not done. Reconnect your cable, admire your handiwork, put on a Kimmie Rhodes CD, power it up, and enjoy your new CD deck!

Larry Cotton is a retired engineer and part-time math teacher who lives in New Bern, N.C., and likes to listen to, write, and play anything musical. He built the player marimba featured in *MAKE, Volume 04, page 20*.

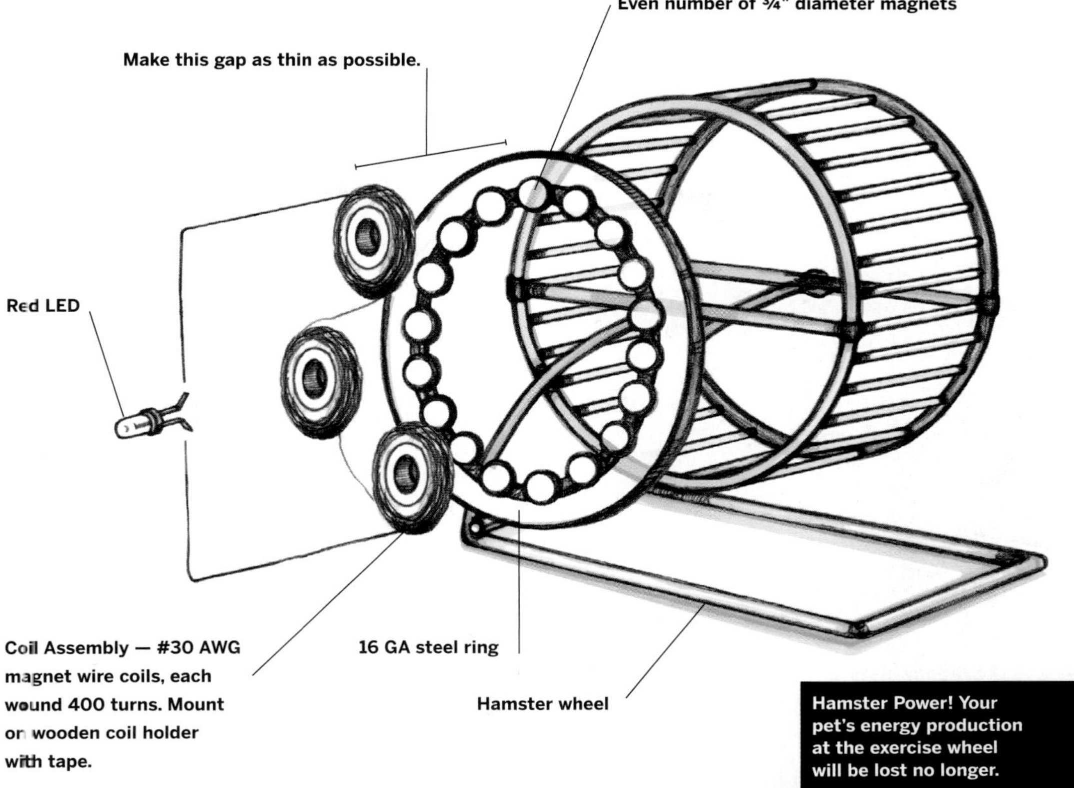

Hamster Power! Your pet's energy production at the exercise wheel will be lost no longer.

RUN, LALA, RUN

A rodent-powered nightlight.

By William Gurstelle

Illustration by Damien Scogin

Which maker hasn't thought, at some time, about putting the pet dog/cat/horse/weasel on a treadmill and actually getting something in return for all the money they've spent on the animal? Well, that time has come. Look out, dear pets, your free ride has come to an end.

The rodent-powered nightlight (RPNL) described here is a custom-made, low-rpm alternator. The original concept came from the folks at otherpower.com, who also sell magnets and wire for the insides of the alternators.

If built well, the RPNL can generate enough voltage to energize an LED when Lala the hamster merrily spins her exercise wheel in the wee hours. According to people who study these things, hamsters and similar rodents typically spin their exercise wheels at about the same rate as an LP record (which opens up the intriguing possibility of a hamster-powered phonograph as well, but that's something for a later issue.)

The Design

My Lala turns an axial flux permanent magnet alternator that has two main parts. First, an even number of strong permanent magnets are arrayed on a steel ring attached to the side of the hamster wheel. This is the Magnet Assembly.

In close proximity to the plane of the magnets goes the second main part: the Coil Assembly. The Coil Assembly consists of two or three bundles of tightly wound wire coils made from 30 AWG magnet wire. As the hamster spins the exercise wheel, the wire coils slice through the magnet's flux lines and generate voltage.

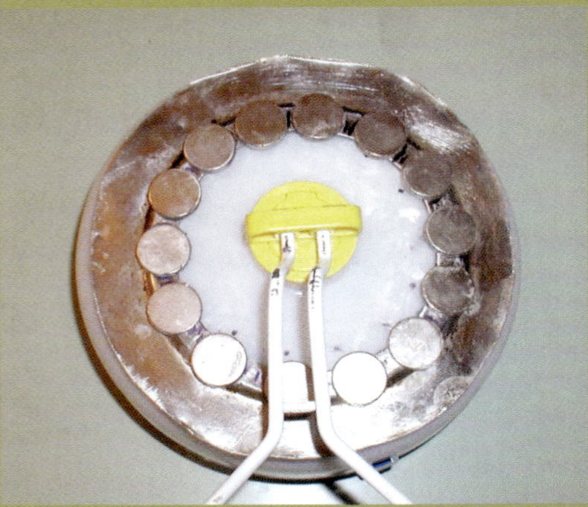

Magnet wheel assembly attached to side of rodent exercise wheel.

The diagram on the previous page shows the basics. Making this nightlight is straightforward, but care is required to make it work. Lala isn't capable of putting out a lot of energy (1 hamster-power = .00016 horsepower, by my calculations), so there isn't much tolerance for inefficiencies.

Step 1: Make the Magnet Assembly

MATERIALS
- 16-gauge (or so) steel ring **Exact ring dimensions depend on your particular exercise wheel.**
- ¾" diameter by ⅛" thick round magnets (18) **You can use more magnets, if they will fit, but there must be an even number of magnets.**

Space 14 of the magnets evenly around the perimeter of the steel ring, forming a perfect circle. The magnets *must* be placed with their north and south poles alternating. Test each magnet before gluing by holding it over its neighbor. They should repel. Then mark out the spacing and glue the magnets in place with Super Glue.

Once the magnets are glued down, use the remaining 4 magnets to attach the magnet ring to the back of the exercise wheel.

Step 2: Make the Coil Assembly

Take a large spool of #30 AWG magnet wire and wind 3 coils by hand. The coils should be tightly wound and thin — around ¼" thick. I used 2 of the plastic separators that came with the magnets as a core (any thin, non-magnetic disk will do) and simply wound 400 turns of wire around the separators, using 2 large washers on a bolt as a wire form. Once the coils are wound, squirt on Super Glue to keep the coils from unwinding.

MATERIALS:
- Spool of #30 AWG magnet wire
- Magnet core material
- Wire winding device **Consisting of one 3"-long, ¼" bolt, 2 fender washers, and 2 nuts**
- Wooden framework **The exact shape depends on your hamster wheel.**
- LED
- Tape

Now, mount the coils to a piece of wood so that each coil is directly over the center of one of the magnets. It doesn't matter which magnet the coil is centered over. What is important is that each coil is directly over a magnet, and the gap between the face of the coil and magnet is as small as possible.

Step 3: Connect Them Together

Once that's done, it's time to connect the wire coils. The best way to do this is experimentally, using a voltmeter. Remove the insulation from the wire ends with sandpaper and connect the wire ends to a voltmeter. Hold the first coil near the wheel and spin it. You should see a voltage reading on the dial.

Connect the second coil to the first coil in series and hold both coils near the spinning magnets. When the voltage doubles, you'll know the coils are wired correctly. Likewise, connect the third coil, and tape all coils to the wooden coil holder you've rigged up.

Connect the 2 loose ends of the wire coils to the LED leads. If you've made the RPNL correctly, a peripatetic hamster will generate enough voltage to light up an LED. For those with lazy or phlegmatic rodents, note that red LEDs require the least voltage to operate.

Now if only we could harness the power of those drinking bird toys with the top hats.

William Gurstelle wrote "Homecasting Digital Music" on page 133.

Photograph by William Gurstelle

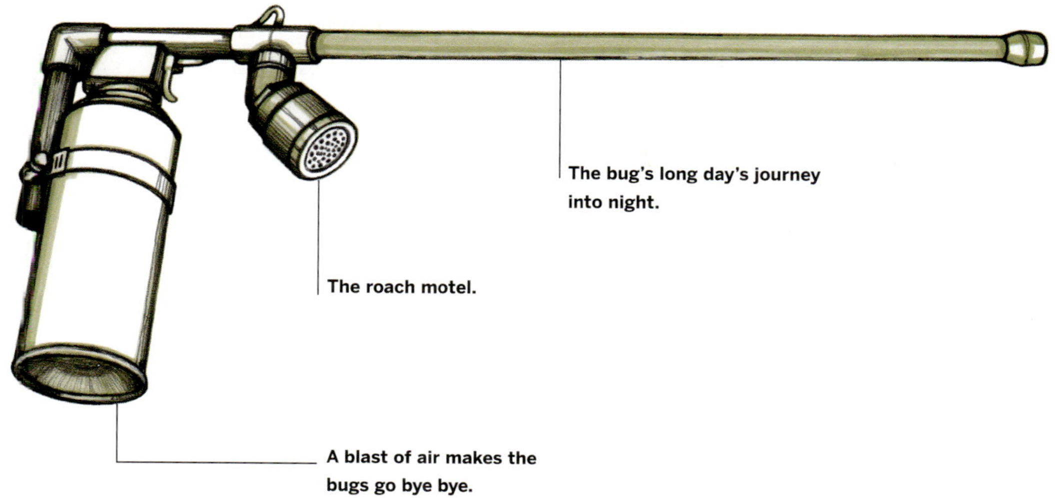

This bug suction device is so much fun to use that you'll relish home spider infestations.

SPIDER RIFLE

Humane, compressed-air-powered bug trapper.

By Matt Lind

I created the spider sucker rifle because my wife and mother-in-law are terrified of spiders in the house. I had seen "bug vacuums" on infomercials but didn't want to shell out the $30-$60 for something that runs on batteries and probably doesn't perform all that well.

What you are about to build is not technically a rifle, as it does not launch a projectile, but it does look like a rifle. It works by directing compressed air downward through the "drop tube," which draws air from the barrel, along with any hapless bug that happens to be next to the barrel end.

Illustration by Damien Scogin

Make It

Important! If you use a saw for cutting the pipe, make sure you wear proper respiratory equipment. This also applies for drilling. Dust from CPVC is toxic and can stay in your lungs. Using a pipe cutter instead of a saw will minimize dust. Check out the materials list on the next page, then begin.

1. Cut the ½" CPVC pipe into 4 lengths: 18" (1), 4" (2) , ¾" (1).

2. On one of the two 4" pieces of pipe, cut a slit lengthwise ½" from one end. The slit should be about ¾" long and go completely through the pipe. This slit will be used to clamp the can of compressed air to the Spider Rifle.

You can either drill a few pilot holes through the pipe and cut the rest out with a utility knife or use my preferred method: a Dremel with a cutting bit. Be careful with a knife, as it can slip easily, and as always, wear a dust mask or respirator.

3. Use a 3/32" drill bit and drill a single hole in the center of the CPVC T-fitting. This will be the

MATERIALS

To construct the Spider Rifle, you will need the following parts:

5' or 10' length of ½" CPVC plumbing pipe $2 for 10 (10' usually costs only 50 cents more than 5')

½" CPVC coupler

½" CPVC T-fitting

½" CPVC 45° elbow fitting

½" CPVC 90° elbow fitting

1"x½" PVC coupler

1" PVC end cap

70mm hose clamp

Can of compressed air $5

Small piece of thin, semi-rigid plastic, such as from a coffee can lid or the walls of a soda bottle

Total approximate cost is $10. This can be more or less depending on the cost of individual components in your area, as well as your choice of adhesive.

OPTIONAL ITEMS

You can use PVC cement, hot glue, or another form of adhesive for connecting parts. I use PVC cement, as it provides a permanent bond.

Combination pack containing 1 can of PVC primer and 1 can of multipurpose PVC cement $4

Can of spray paint in color of your choice $1

Can of spray clear coat (protects paint) $1

hole that the straw from the can of compressed air goes into. You can use other sizes of drill bits, but you may have to ream it out a little if you use a smaller one or fill the gaps with hot glue if you use a larger one. Any type of drill should work, but a drill press is best. Since you are working with a fitting, you can connect a piece of scrap pipe into the end, to give you a handle to hold on to.

4. Drill as many holes as possible into the 1" end cap. The more the better, to let air through.

5. Trace and cut out the T-fitting stop valve. This can be cut out of pretty much any semi-rigid plastic. I prefer coffee can lids, as they are thick enough, easy to trace on, and easy to cut.

First, take a piece of ½" pipe, stand it on top of the plastic, and trace around it using a marker such as a Sharpie.

Then cut out the circle with scissors and place it into the T-fitting. If it doesn't slide in easily, cut small amounts off until it does. Be careful not to cut too much off or it will slide through. You want the stop to rest up against the ridge near the middle.

6. Push the 4" piece of pipe without the slit firmly into the side of the T-fitting that contains the plastic stop.

7. Connect the 90° elbow to the end of the 4" pipe.

8. Push the ¾" pipe into the bottom of the T-fitting.

9. Connect the 45° fitting to the ¾" pipe as shown in the illustration on the previous page (it's used to attach the "roach motel").

10. Connect the other 4" pipe to the 90° elbow fitting.

11. Connect the ½" to the 1" coupler to the end of the 45° elbow.

12. Connect the barrel to the T-fitting.

13. Connect the ½" coupler to the end of the barrel.

14. Prime and cement all the joints, or use the adhesive of your choice.

15. Cut the ½" coupler at the end of the barrel in half. There is usually a seam on the coupler that you can use a guide.

16. Insert the clamp through the handle and partially tighten.

17. Twist the 1" end cap onto the bottom of the 1" coupler.

18. At this point, you can paint the rifle if you so choose. Allow the paint to dry thoroughly.

19. Insert the plastic air tube that came with the can of compressed air into the top of the T-fitting.

20. Slide the can of compressed air up through the clamp so that the top rests firmly against the rifle. Tighten the clamp so that it is snug. Don't overtighten, as the can is under pressure. Connect the air tube to the can of air.

To use the rifle, simply point the barrel near the insect and squeeze the trigger. The air is forced through the bottom of the drilled PVC end cap. Since the T-fitting is blocked on the side connected to the handle, by forcing air downward and out the bottom a vacuum is created in the barrel. This vacuum draws the insect inward and into the "catch" (drilled 1" end cap). Build and use this at your own risk. But most importantly, enjoy your soon-to-be spider-free home!

See an animation of the Spider Rifle in action here: mslworks.com/SR1.gif

---

Matt Lind is a frustrated consumer living in Appleton, Wis., who has always enjoyed thinking of new uses for the parts in existing products.

Recursion Alert: View this image through red-cyan glasses to see how to take a 3D photo, in 3D.

3D Photography

Taking stereographs is easier and more fun than ever. By Bill Coderre

3D photography has been around for almost as long as regular photography. But the 1950s brought the fad of 3D horror movies, starting with *Bwana Devil* (1952) and *House of Wax* (1953). Many of these were so bad that 3D soon became an embarrassment. Even 50 years later, 3D still hasn't shaken its reputation as a cheap gimmick that causes headaches — which is a shame, because the medium can produce compelling images worthy of contemplation.

It's time for a 3D renaissance. In the last few years, computers have made it simple to produce color 3D images in just a few minutes. All it takes is a digital camera, some free software, and a pair of red-cyan glasses. I'll explain how to do this, and also suggest some camera projects that can make taking 3D pictures even easier.

Photograph by Bill Coderre

If you don't have red-cyan glasses, you can find some at a comics shop, attached to any Ray Zone title. Or, you can buy them from American Paper Optics (*see page 145, "Further References"*).

The Stereo Shuffle: Taking Basic 3D Photos

Grab your digital camera, and head outside. Find a scene with a fun variety of distances to focus on. The closest objects should be at least 5 feet away, and perfectly still. Also, make sure that everything in the scene will be in sharp focus. Sunny conditions will help your depth of field.

Ready? Hold the camera to your eye, and lean very slightly to the left. Snap a picture. Lean very slightly to the right, and snap another picture. That's your basic "stereo shuffle" right there.

Stereo 3D Software

A big list of 3D software is at anabuilder.free.fr/Concurrents.html

Java **(runs on Mac, Windows, Linux, and probably your toaster)**
AnaBuilder anabuilder.free.fr/indexEN.html

Windows
Anaglyph Maker stereoeye.jp/software/index_e.html
StereoPhoto Maker stereo.jpn.org/eng/stphmkr
3D Maker tabberer.com/sandyknoll/more/3dmaker/3dmaker.html

Mac
StereoPress (runs in Classic mode) makezine.com/go/stereopress
Anaglyph Maker lamarchefamily.net/nakedsoft
3D Maker tabberer.com/sandyknoll/more/3dmaker/3dmaker.html

For best results, make sure that the camera moves from left to right about 2 inches, and does not move up or down, toward or away from the scene, or tilt in any direction. Also, do not realign the frontmost object to the same position in the frame. (For the nerds: no y or z translation; no roll, pitch, or yaw; and do not "toe in.")

Now head back to your computer, import the pictures from your camera, grab some free software (see table above), and use it to load the left and right frames of the pair. I recommend AnaBuilder, which has all the functionality you'll need, once you figure out the buttons and menus.

Display the picture as a color anaglyphic image, and before you put the glasses on, correct any twist or vertical offset. The horizontal lines in both views should overlap and be parallel, and objects in each image should occupy the same vertical position in the frame.

One last step: make sure that the frontmost object in the scene has its two views completely superimposed, with no red and blue halos around it. This means that the nearest object in the picture will be at the stereo window depth, not sticking out in front of the frame. This makes the scene easier for most people to view. You can experiment with "eye-poking" stereo later.

Now, put on the glasses and admire your handiwork. Believe it or not, that's it for the basic method. Everything from here on is just enhancements and alternatives.

Near or Far Subjects: Adjust the Interaxial Distance

If your subject is a small flower or a mountain, having it 5 feet away won't work. With a close shot, 2 inches between pictures (called the "interaxial distance") is too much, and the picture will be unviewable, or at least headache-inducing. With a distant subject, 2 inches will not be enough to give any appreciable 3D effect.

For these situations, the basic rule is to divide the distance to the nearest object by 30, and move the camera that amount between shots. That means 176 feet for an object a mile away, and 0.8 inches for an object that's 2 feet away.

These approximations, however, do not take into account important factors such as lens focal length and the distance of the farthest object. For a more rigorous approach, see the Bercovitz Formula article listed under "*Further References*."

More Accurate Alignment: Make a Candy-Tin Slide Bar

A stereo slide bar will ensure that your camera moves only horizontally between pictures. You can easily make one out of a tripod, a bubble level, and a candy tin.

Use a nibbler, a drill, or some other tool to cut a slot in the bottom of the tin. The slot should be a bit wider than ¼", running the full length of the box from side to side, parallel to the hinges, centered halfway between the hinge side and the front. Place the slot onto the ¼-20 head screw (¼" diameter, 20 threads per inch) that sticks up from the top of your tripod, and align the candy tin so that it is parallel to the tripod's handle. Use poster putty (such as Fun-Tak) to attach the camera to the candy tin, and the bubble level to either the camera or the tripod. To take pictures, make sure the tripod is level, then snap one photo, slide the camera over, and snap the second photo. It also helps to add a ruler to the back of the candy tin, so you can control your interaxial distance.

Live Subjects: Link Two Cameras

You can take 3D photographs of subjects that don't hold still by using two cameras. Place them next to each other on a bar, or tack them together by their bottom surfaces and rotate or invert the images so they match later. If the lenses cannot be placed close enough together, you can rig up a mirror system. The other half of the solution is finding a way to trigger both shutters at the same time. This depends on the type of cameras you're using (*see "Further References"*).

"Found" 3D

My friend Eric von Bayer pointed out that you can create stereo pairs by selecting frames from 2D movies. Copyright considerations prevent me from showing this example, but you can download the *Star Wars Episode I* trailer, and extract two frames from the shot that pans horizontally over a landscape of Naboo. Drop them into AnaBuilder, and you end up with a pretty nice 3D scene.

Loreo Cameras

Loreo (loreo.com) makes inexpensive 3D cameras that use mirrors to put the left and right stereo images on one standard 35mm film frame. For viewing, you could either use their included viewer, or get a digital image, and then cut and paste the two sides into the software provided. The Loreo doesn't let you adjust the exposure, but at least it has a built-in flash.

Other Display Options

Red-cyan glasses take some getting used to, and you can see "ghosts" when the disparity between the images is large. Here are some other display technologies.

The **ColorCode** system substitutes blue and amber for red and cyan. This reduces ghosting, but it's also patented, and illegal to distribute software that produces ColorCode images.

Adjacent-image pairs are viewable by crossing or uncrossing your eyes, but this is a tricky skill to learn, and it limits image size. To view bigger images, you need a stereoscope, like the ones from the turn of the 20th century. These devices have two lenses mounted off-center, to allow for close-focus, and a prism that lets the eyes point straight ahead while the view-lines diverge. Fancier viewers use mirrors to accommodate larger stereo pairs. You can make your own stereoscope, but note that the brain usually interprets misaligned left and right eye views as caused by food poisoning, and reacts by the sudden expulsion of the contents of the stomach. Therefore, it's important to align the mirrors precisely.

Time-interlaced images are viewed by LCD shutter glasses, which you can get for $30 from Razor 3D (razor3donline.com). Unfortunately, you cannot use these glasses with LCD monitors, progressive scan monitors, or any monitors set to refresh faster than 90Hz. As a result, images often seem to flicker, but it helps to turn down the brightness of the monitor, and reduce room lighting. At standard television frequencies (roughly 50-60Hz), movements appear jerky, and fast-moving objects will often "tear."

Horizontally interlaced image sets are displayed by putting the interlaced image behind a lenticular screen, like with 3D postcards. Free software such as Interlace produces interlaced images that you can print on a high-resolution printer, and then display in a sleeve behind a reusable lenticular screen. The big plus of this technology is that it does not require the user to wear glasses or use a stereo viewer. The downsides include low horizontal resolution, and the difficulty of aligning the image with the screen.

Several television companies have hinted at future **3D TV** products based on lenticular screens, but the NTSC format's width of 720 isn't enough to split in half and retain a good picture. With horizontal resolutions of 1,280 and 1,920, HDTV seems like the "silver bullet" that might make it all work. We can only hope.

Further References

General 3D information: stereoscopy.com and dddesign.com/3dbydan/3dlinks

Explanation of the Bercovitz Formula: makezine.com/go/vic3d

Synchronizing twin cameras: makezine.com/go/twincam

American Paper Optics, manufacturer of 3D glasses and experimenter kits: americanpaperoptics.com

Ray Zone, "The King of 3-D Comics": ray3dzone.com

Bill Coderre dredged most of this article from class notes and memories of a seminar by 3D pioneer Stephen A. Benton, professor at the MIT Media Laboratory, to whose memory this article is dedicated. These days, Bill is a programmer for Apple.

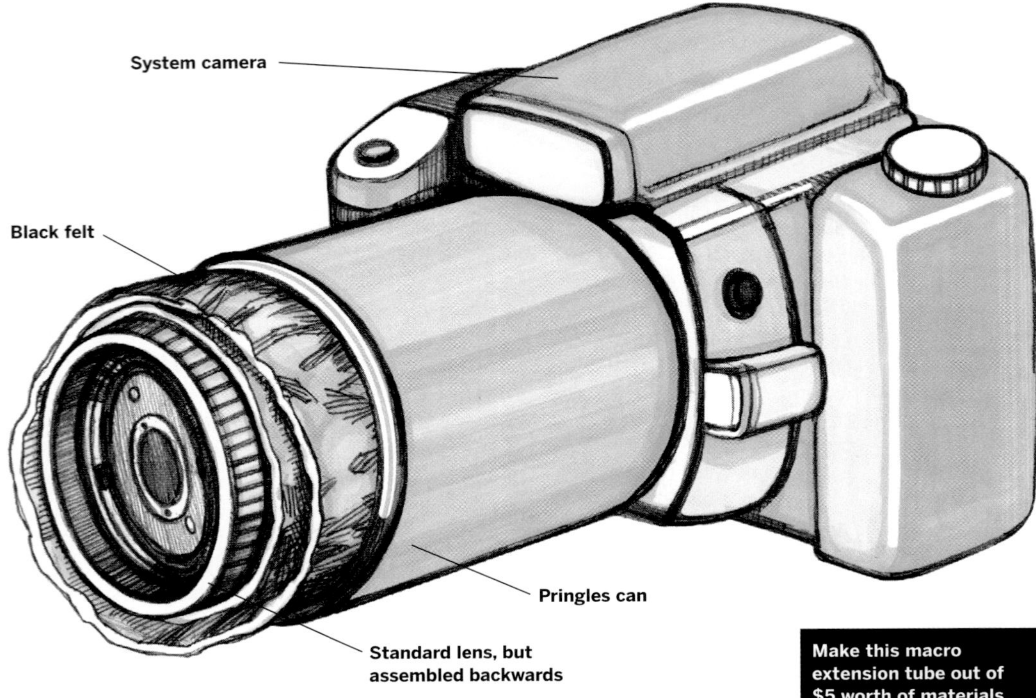

Make this macro extension tube out of $5 worth of materials.

Macro Photography On a Budget

Pringles-can lens extender produces dazzling ultra close-ups for peanuts. By Haje Jan Kamps

Photography is often seen as an expensive hobby, but here's a way to create a lot of fun for very little money: a macro extension tube you can make for less than $5 worth of materials.

Macro Tube Theory

The concept of a macro tube is as old as photography itself. Normally, a lens takes incoming parallel light and focuses it down onto the film or imaging chip. By turning the lens the wrong way around, the opposite happens: light focused in on a small area is refracted to become more parallel.

With the lens in its wrong-way configuration, either by using a reversal ring or by just holding it against your camera body, you can take photos up close. In theory, the farther away from the camera you hold your lens, the higher magnification you get. The problem is that the "air" between the camera and lens element can't have any light leaking in. Furthermore, there cannot be any lens movement when you take photos, as this will result in blurry photos. The solution? A macro extension tube, which holds the lens and shuts out light.

Extension tubes and bellows can be bought, but that's definitely more expensive and probably a lot less gratifying than building something yourself. For our tube, we'll use a Pringles can,

Illustration by Damien Scogin

which is long and lightproof. At one end, we'll attach a camera body cap with a hole cut out of the middle, which will make it easy to screw the extension tube into the camera body. At the other end, we'll put the reversed lens.

MATERIALS

System camera with removable lenses (SLR or similar)
Body cap for your camera
Pringles can, or tube of similar dimensions
Standard (i.e. non-zoom) lens
Dremel or other cutting tool
Black felt
Sandpaper
Hot glue or epoxy

WARNING: Throughout this project, the insides of the camera are exposed to dust and dirt, and cameras hate both. It may be best to do this experiment with an old, film-based SLR. If you wind up having to take your Digital SLR camera in for servicing, it will be far more expensive than just buying a macro lens in the first place.

Perforate the Cap

Use the Dremel to cut a hole in the camera body cap. Cut out as much as you can, but make sure to leave the whole bayonet mount intact. Sand the inside edge to matte the surface; this edge may be subjected to light from the lens, and you want to prevent reflections.

Prepare the Tube

Mark the hole in the body cap on the bottom metal end of a Pringles tube, and use a Dremel to cut a neat hole in the bottom of the tube. When this is completed, make sure to completely clean the inside of the tube — you don't want aluminum shavings (or salt and grease) in your camera body. Then line the inside of the tube completely in black felt. This will stop reflections, which will dramatically improve the photo quality.

Attach the Cap

Glue the body cap to the Pringles tube using hot glue or strong epoxy. This connection has to bear quite a bit of weight, so use strong glue. If the glue dries translucent, use black paint or a felt-tip marker to darken it and prevent light leaks.

Choose and Install the Lens

Most lenses, including zoom lenses, should be perfectly usable for this project, but I recommend a standard lens, as they are usually cheaper, sharper, and easier to focus. I used a Canon 50mm f/1.8 MK2. It is the cheapest lens in Canon's catalog, but also one of the best.

If you're lucky, you'll be able to slip the lens inside the open end of the Pringles tube, and it will fit snugly against the felt. Failing this, wrap another layer of felt around the lens, tape it down, and try again. Be sure to put the lens into the tube backwards, with its electrical contacts facing away from the camera body.

Take Photos!

You don't focus this lens by using the focusing rings; instead, you just move either the camera or your subject matter nearer or farther. This lens has next to no depth of field, so focusing in this way can be frustrating.

For setting exposure, your camera's internal light meter is actually a good starting point — it isn't always accurate, but it will give you an idea. Be careful, though, because any movement during the exposure will blur the image. To prevent blurring, use the self-timer or a remote shutter lead. If your camera supports mirror lock-up, use this as well, to further reduce vibration.

Taking good macro photos takes a lot of practice, and it's often a technical exercise rather than a creative one. But it's fun, and it opens up a world of photos that would otherwise have been impossible. Experiment, create, and enjoy!

Canon EF Aperture Hacking

All of Canon's newer lenses have electronically controlled aperture, so you can't normally stop down a lens unless it is attached to the camera. There is a trick you can use, however: put the lens on the camera, and set the camera to fully manual (M). Select the aperture you want. Then, press and hold the aperture preview button, and take the lens off the camera. If you have done it right, you are now holding a lens that is still stopped down at the aperture you selected.

Haje Jan Kamps is a freelance writer and photographer based in Bristol, England. He runs photocritic.org for DIY photographers and their projects.

An easy alternative to insanely overpriced digital picture frames.

$130 Digital Picture Frame

Build one in 15 minutes.

By Mike Kuniavsky

Call me a cheapskate, but $400 for a digital picture frame when a complete laptop costs $600? I won't stand for it!

Call me lazy, but a week of evenings with a Torx screwdriver, a razor blade, and an old laptop, trying to get a memory-resident Linux install to boot? I have better stuff to do!

I've wanted a digital photo frame for a long time. Unfortunately, my cheapness and laziness have stood in the way. Paying $400 for an LCD and some driver hardware (when you can get a DVD player for $85) just seems wrong. And hacking apart an old laptop is the kind of project that invariably gets put off indefinitely.

Until recently, I figured I'd end up waiting for the commercial ones to get cheap, but then I stumbled on a solution that satisfied the lazy cheapskate in me. Here's my $130, 15-minute solution.

Assemble

The fast version:

Plug everything together, and stick it all into a picture frame.

The slightly slower version:

1. Unpack everything (1 minute).
2. Plug the audio and video cables together. The yellow cable is the video cable, and is the most important. Plug the red and white ones together if you want to listen to MP3s or hear the sound of video (1 minute).
3. Copy some photos from your PC to the memory card, plug it in, and turn everything on to make

Photography by Mike Kuniavsky

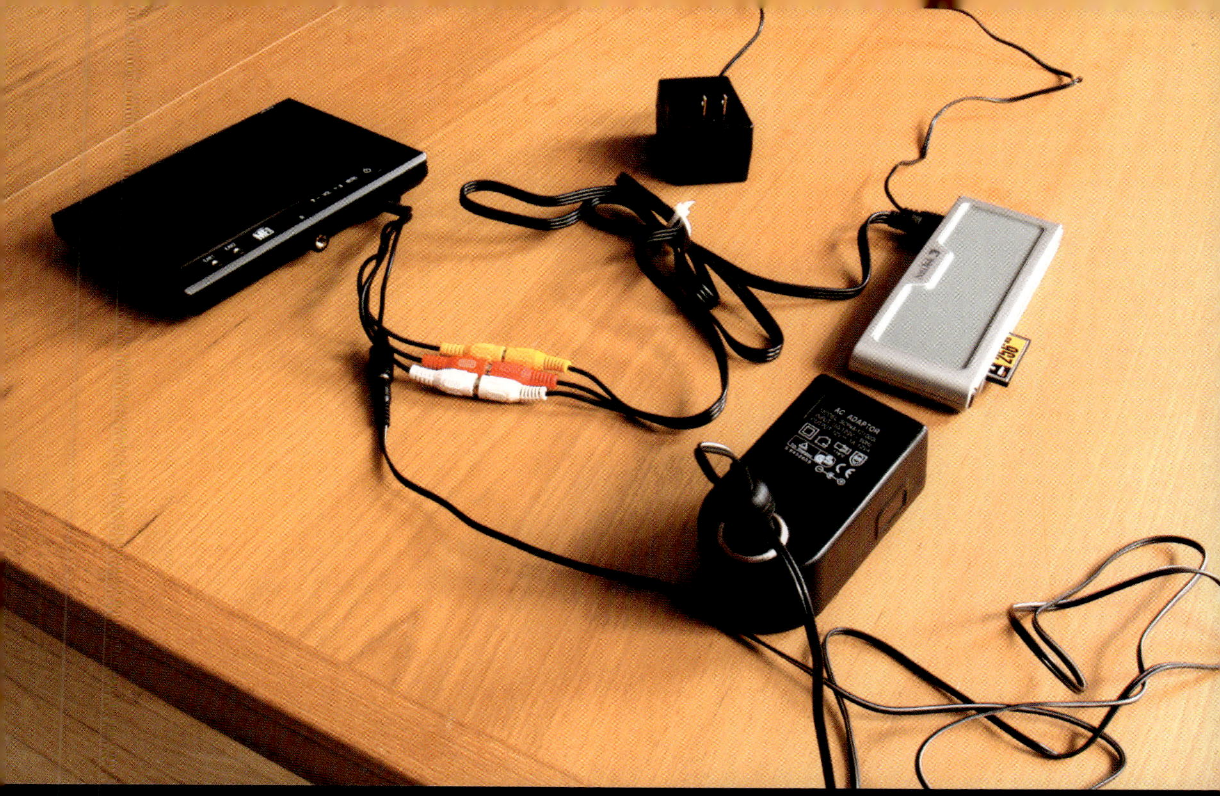

Photos are stored on a memory card that's plugged into a digital photo album/solid-state video player. These are small devices that let you watch photos and video from your digital camera on a TV screen. The player feeds a car video display of the type designed to keep kids quiet in the back seat. The display is powered by an adapter that converts house current into 12-volt, cigarette-lighter DC.

MATERIALS

Headrest video display for cars These are available on eBay for between $50 and $200, depending on the size, the brand, and whether you get them singly or in packs of two (one for each headrest). Avoid monitors that have extra hardware. Monitors already embedded in headrests or that come with dash-board/ceiling mounts have to be extracted, which, you know, takes time. Make sure it has an RCA video input jack.

Memory-based video player Solid-state video players are likely to be smaller, cheaper, and more reliable than hard-disk or DVD players. I tried the SanDisk Digital Photo Album and the Neuros MPEG-4 recorder for this one, but there are others. These run between $30 and $100.

Auto accessory wall adapter Essentially 12V transformers with a car cigar lighter plug, these let you recharge your mobile phone using your car charger, at home. I don't know why you'd want this, but someone clearly does, and these are on the market. I found one with 1 amp of output power, which is more than enough to drive a big LCD without breaking a sweat, for $3.50 at Electronic Goldmine.

Picture frame from a thrift store I was lucky to find one for about $2 that was an old perfume box/picture frame, just the right size to fit the monitor and the player. The box-style frame is nice because the LCD screen, the player, and the data and power cords all fit inside. Otherwise, you might have to build a box to hold all the stuff neatly. And that would take energy and money.

A memory card You'll want a dedicated one, since it's going to sit in the player most of the time, and the bigger the card, the more photos you can put on it.

Tape I used clear packing tape because it was handy, though any kind of wide tape will work.

sure that all the electronics work. If everything works, take it apart to install all the stuff in the frame (2 minutes).

4. Install into the frame. This step will probably take the longest. In my case, I stuck the components into the box with zip ties and tape, but you could build an armature, cut a mat for the frame, and permanently affix the whole thing inside a frame with glass. I'm too lazy to do that, but you go for it (9 minutes).
5. Copy more photos to the memory card (2 minutes).
6. Plug it in (0 minutes)!

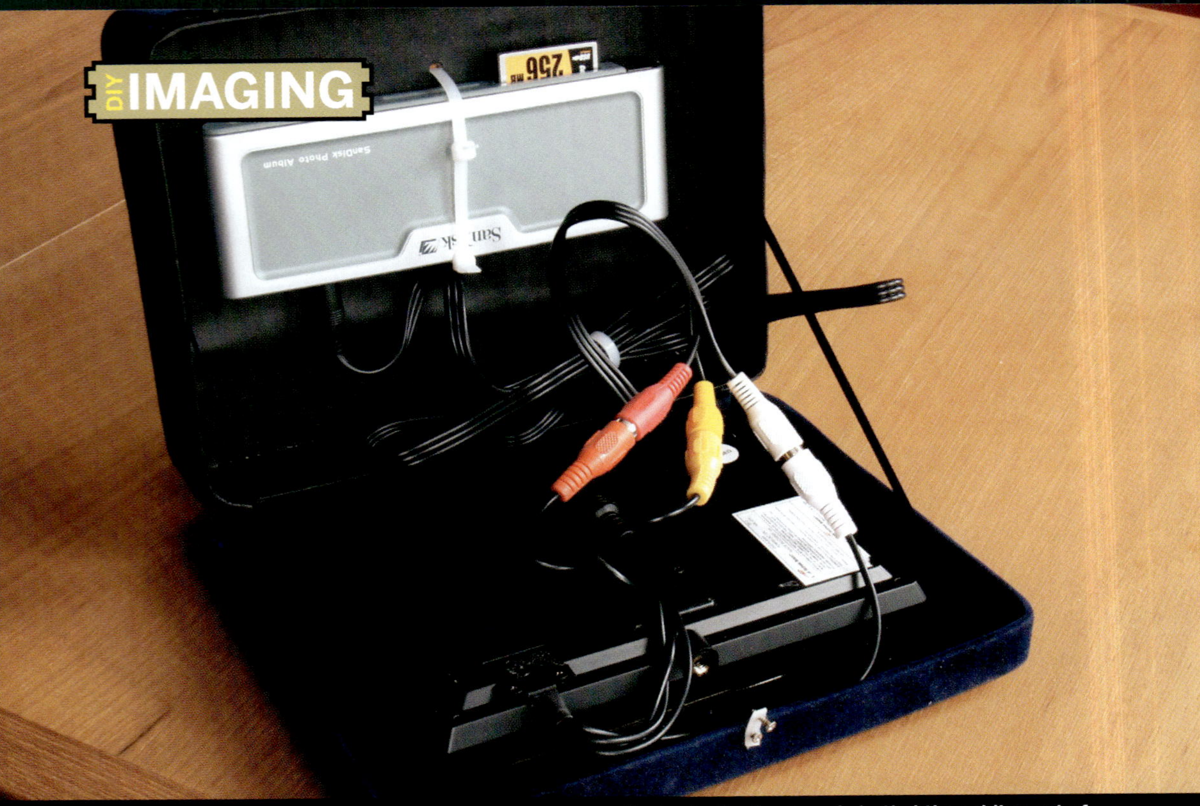

Components tucked inside the perfume box/picture frame. I zip-tied the player to the back, taped the display to the front, and zip-tied the cabling out of the way.

Things to remember:

1. Leave enough room to remove the memory card, so that you can add more photos later.
2. Don't entirely cover the infrared LEDs, so you can use the remote controls for the monitor and the player when it's assembled.
3. Adjust the screen to your liking before you cut the mat. LCD display proportions are 16:9, but most digital photos are 4:3. This creates a dilemma. Here are your three choices:

3a. Stretch pictures to fill the whole screen horizontally, like many people do with their TVs. This makes everyone in your pictures look 20% shorter and fatter.

3b. Leave black bars of unused screen to the left and right of the picture. You can mask these out with an appropriately cut mat.

3c. Clip the top and bottom of the images so they fill the full width of the screen without distortion.

I prefer not distorting or cropping my photos, so I chose 3b. Both the player firmware and the monitor setup menu let you configure this.

More Stuff to Do

These memory-based players can play back video (that's what the Neuros is best at; the SanDisk has better photo slide show functions), which adds a whole other dimension to what they can do. In addition to photos, I've been taking short video clips for the last year with my digital camera. The quality isn't as good as a video camera, and the players are finicky with the file formats they'll play, but slide shows of 10- to 30-second clips are surprisingly engaging, while still being ambient.

Of course, if you converted a full-length video to the appropriate format and copied it onto a memory card, you could watch a feature film on your picture frame. But why do that when that's what TVs are for!

My next step is to set up a script to automatically download the latest photos from my friends' Flickr streams onto a memory card, and then, maybe once a week, I'll swap cards in the frame — if I remember.

Sometimes being cheap and lazy is just so much work!

Mike Kuniavsky is a San Francisco- and Portland-based ubiquitous computing and user-experience consultant and writer who blogs at orangecone.com.

THE SECRETS OF MONITORING ATMOSPHERIC HAZE

MAKE A MEASURING INSTRUMENT WITH AN OLD VIDEO CASE AND $20 WORTH OF PARTS.

By Dr. Shawn

EVEN ON THE CLEAREST OF DAYS, THE atmosphere over our heads buoys up thousands of tons of nature's crud — pollen, mold spores, microscopic fragments of decayed plants, insect parts, wind-borne dust, water vapor, smoke from forest fires, and volcanic ash — that we collectively call haze. Of course, we humans contribute more than our share as well. Smog is the best-known example of artificial haze.

Although many cities monitor the quality of their local atmosphere, shockingly little is known about how haze is changing globally because, believe it or not, no one is coordinating haze observations around the world. Considering how important these observations could be for the health of all life on our planet, this lack of attention is not only shameful, it could also be downright dangerous.

All this may soon change, thanks to the inventiveness of one of America's greatest living citizen scientists. Forrest M. Mims III is renowned as a science writer for mentoring more than a million professional and amateur gadgeteers through a series of wonderfully earthy, hand-drawn how-to booklets that he wrote for RadioShack. And even though he has no formal scientific training, Mims is a living legend amongst amateur scientists for his knack at besting the professionals at their own game.

In 1991, for instance, while standing in his backyard testing a $200 hand-held device of his own design, Mims discovered that an instrument costing perhaps a thousand times more than his, flying aboard the Nimbus 7 satellite, was steadily drifting out of calibration. That highly sensitive spectrophotometer, developed by NASA's elite corps of scientists and engineers, was the very same that the space agency had used to discover the "ozone hole" (really, a roughly 50% reduction in ozone concentration) that appears during the coldest winter months in the stratosphere above the Antarctic. Indeed, back then, this spectrometer was NASA's principle source of information about the health of the protective sheath of molecules that block the sun's most damaging ultraviolet rays.

NASA was understandably reluctant to admit that it had been shown up by an amateur. However, to their credit, when the red-faced researchers finally did own up to the error, they made Mims a collaborator. NASA sent this dogged citizen scientist to Brazil, Hawaii, and the western United States to study the atmospheric impacts of biomass burning and volcanic eruptions using his clever collection of low-cost homemade instruments.

MIMS' VISIBLE HAZE PHOTOMETER

One of Mims' ingenious gadgets is his Visible Haze Photometer. This remarkable device can be cobbled together from an old videocassette case and about $20 worth of RadioShack parts. Moreover, it's so simple to make that even the most hardened technophobe can put it together in an afternoon. This instrument could revolutionize haze research by opening the field to all comers: citizen scientists, science fair students, and weather watchers of all stripes.

The Visible Haze Photometer exploits the fact that the sun is the perfect probe for measuring haze. The intensity of sunlight striking the top of the Earth's atmosphere is essentially constant. So, by measuring the intensity of the sunlight at the ground and knowing the thickness of the atmosphere it has passed through, one can determine how much light has been either scattered or absorbed and then determine how much gunk is floating about overhead.

There is a complication, however. Air molecules themselves scatter light, a phenomenon known as Rayleigh scattering. The effect may create beautiful blue skies and brilliant red sunsets, but the fact that air molecules scatter blue light more strongly than red is a pain when it comes to measuring haze. To compensate, one must look only at a narrow sliver of the spectrum over which the scattering is constant.

To achieve this selectivity, professionally designed instruments typically started with a detector that could see the entire visible spectrum, and then added an expensive optical filter that permitted only a narrow band of frequencies to pass. But with no government grants to rely on, Mims needed to find a cheaper way.

LEDS: A CHEAP AND SIMPLE SOLUTION

His solution was to use a miracle of modern technology, called a light-emitting diode (LED), that costs just pennies. Electronically, LEDs couldn't be simpler. Just dump current into the bottom, and light of a single color flows out the top. Mims knew that this, like most natural processes, could run backward — light falling on an LED could create a current that could then be detected. Furthermore, just as LEDs produce a single color, an LED generates a current only when light of just the right color strikes it.

Mims' photometer uses a green LED that emits light at around 555 nanometers and detects light at around 525 nanometers. RadioShack sells them for under a dollar. The circuit contains just two other active elements: an operational amplifier and a resistor. Mims houses his instrument inside a black plastic VHS videocassette case with a hole drilled in one end to admit sunlight.

An LED can only detect light that has a shorter wavelength than the light it emits. Photons are emitted in an LED from a very special set of energy states that atomic physicists call the "conduction band." These are basically states in which an electron is free to wander about the substance. When emitting light, the energy to pump the electrons into the conduction band is supplied by current flowing in through the leads. It turns out that, to be detected, a photon has to supply this extra energy itself. And since photons with shorter wavelengths have more energy than photons with longer wavelengths, an LED's detectible photons always have shorter wavelengths than its emitted photons.

MAKE A PHOTOMETER

1 Prepare the LED

The body of the LED is epoxy. The top is shaped like a lens to concentrate the LED's relatively dim light into a narrow cone that is turned toward the viewer. It turns out that more of the sun's light reaches the sensor when the lens and most of the LED's body is removed. Coarse sandpaper and a little elbow grease do the job fast. File the case down vigorously until you've whittled it to within about 4mm of the light-producing element at the bottom.

2 Wire and Install the Circuit

Now, build the circuit and install the solderless breadboard in the VHS case. Keep a big gap between the LED and the hole that you will eventually drill in the top of the case to make sure that only the sun's light — and not the ambient light in your surroundings — reaches your detector. (You can use a hot glue gun if you're making this instrument for a short-term project, like a science fair. If you intend to get serious, then use tiny ¼" machine bolts to lock the breadboard into place.) Next, install the terminals, the battery holders, and the indicator light into the case where shown.

3 Test the Circuit

It's time to test your circuit. Indoor electric lights are too feeble to produce much signal, so take your instrument outside. With a multimeter, record the so-called "dark voltage" — the voltage you measure when the case is closed. With Mims' choice of components, it's likely to be rather large — perhaps several hundred millivolts. Next, open the case and let the sun's light shine down directly onto the circuit board. If the signal jumps several volts when you open and shut the case, your circuit is working.

4 Secure the LED

Use one corner brace to keep your LED sensor from shifting about over time. Slip the LED through the

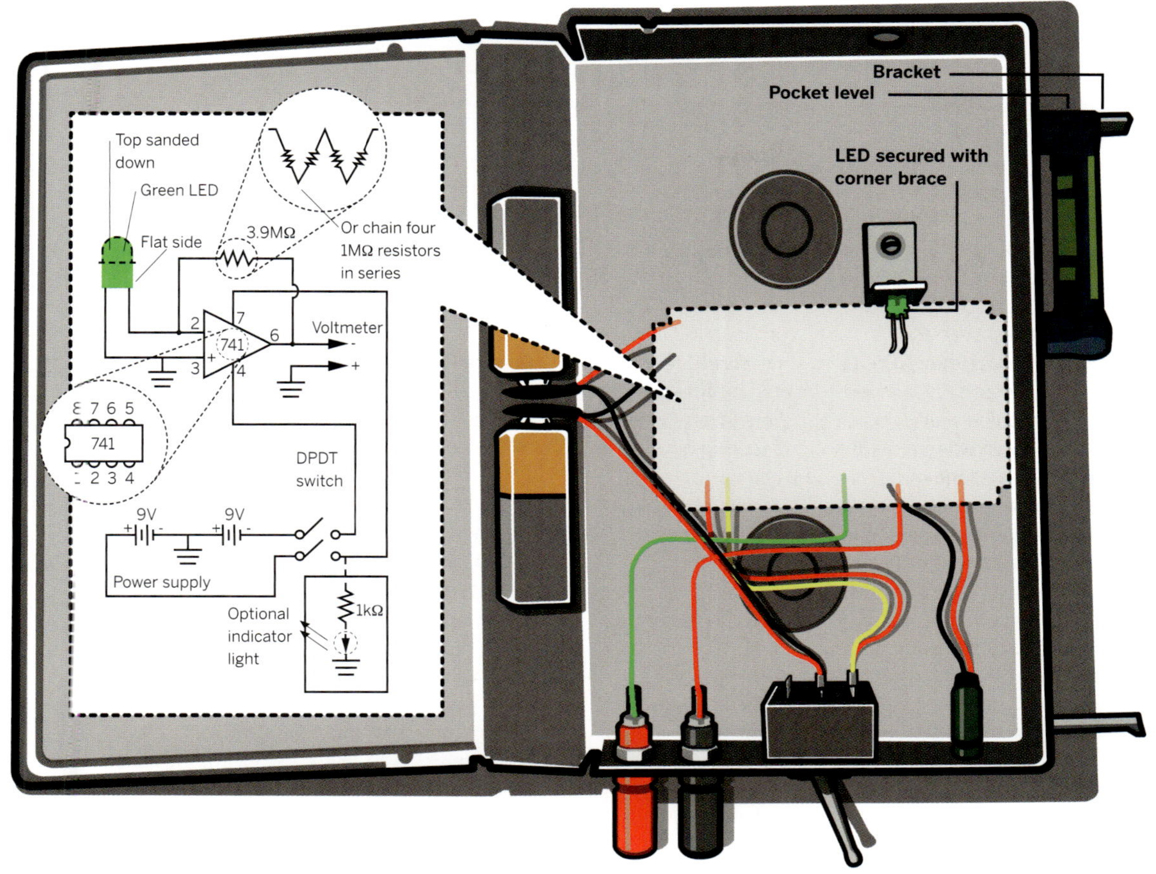

Illustration by Damien Scogin

Make Your Own Haze Photometer

Here are the parts you'll need:

- **741 op-amp (RadioShack 276-007)**
- **3.9M ohm resistor** (RadioShack does not happen to sell resistors with this value), **or four 1M ohm resis tors** connected together end-to-end to form a chain (271-1130). Mims' circuit calls for a 3.9M ohm resistor because this is a standard value. But linking four 1M ohm resistors together to create a 4M ohm resistor will work just fine.
- **Small solderless breadboard** (276-175). This device eliminates the need to solder wires together and makes building the circuit a snap.
- **Green LED** (276-022A)
- **9V batteries** (2)
- **9V battery caps** (2) (270-0325)
- **Hookup wire or, much better, a wire kit** (276-173), which is a small plastic case containing pre-stripped wires of various lengths. Wire kits are a great time and aggravation saver.
- **Double-pole double-throw switch (DPDT)**, since it can turn off both the positive and negative power to the op-amp. RadioShack's model 275-652 is perfect.
- **1" (25mm) L-shaped corner braces** (3) from your local hardware store.
- **Black (and therefore, opaque) VHS case**
- **Pocket level**

Optional, but strongly recommended:

- **Battery clips or holders** (2) (270-326B). These hold the batteries in place inside the case. You could always use duct tape, but holders make fieldwork easier.
- **Banana plug terminals** (2) (274-662). You could feed the signal to your voltmeter by passing two wires through the case, but terminals make data taking more convenient.
- If you decide to use banana plug terminals, then you'll also need **banana plug connectors** (2) (274-721) to go with them.
- I've added an **LED indicator light** (276-309) to Mims' original design so I can tell at a glance when the circuit is on. If you want to add it to your instrument, you'll also need a 5kOhm resistor (271-1124) to limit the current in the LED.

hole in one side of the brace as shown, butt it firmly against the breadboard, and either hot glue, bolt down, or epoxy the brace in place.

Next, you'll need to drill the hole in the top of the case that lets in the sun's light. Carefully position it *directly above* the LED detector. The hole should be roughly the same size as the hole in the corner brace; 6" is just about perfect.

5 Mount Corner Braces

The other two corner braces, when attached to the outside of the case, will enable you to both align the instrument with the sun and measure the all-important sun angle — the angle between the sun and the horizon. If you'd like to be able to make measurements even when the sun is low on the horizon, you'll want to install the braces as far apart as possible. But make sure the vertical tabs face each other so that a paper ruler can be pasted flat between them. Hot glue, screw down, or epoxy the tabs firmly in place.

6 Make Alignment and Sun Angle Gauges

Next, place a small piece of white tape over the vertical tab on the bracket nearer the middle of the case. Then open the case and point the instrument toward the sun. Move the case around until the bright spot of sunlight that falls through the hole you drilled is centered on the LED sensor. If the brackets are approximately aligned, a second image of the sun, this one caused by the sunlight passing through the hole in the front brace, will appear somewhere on the tape. Lightly mark its center in pencil. The case may flex slightly when closed, so you may need to adjust your mark. Close the case and align the instrument to the sun using the outside brackets. Then tilt the instrument slightly in different directions while watching the voltmeter until you find the orientation that produces the largest voltage. Make a permanent mark at the center of the sun's image. From then on, whenever you center the bright spot on that mark, you'll know that your sensor inside the case is fully bathed in sunlight.

Next, paste a paper ruler on the case between the two brackets. You can photocopy a metric ruler. Finally, glue a pocket level to the case. Pocket levels work well, or you can purchase a small line level and, using a pair of pliers, break off the plastic parts used to connect the level to a string. With the level attached, you're ready to go!

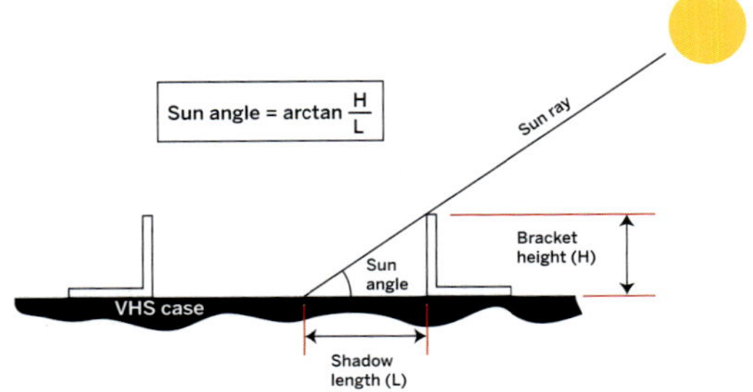

Calculate the sun angle by dividing the height of the bracket by the shadow length of the bracket. The arctangent of this value is the sun angle.

TAKE A HAZE MEASUREMENT

To make a measurement, you first record the dark voltage, that is, the voltage produced when no light strikes the detector. Cover the hole in the top of the case and record the voltage. Now, center the image of the sun on the mark you made on the lower brace and record the resulting voltage. The signal is this voltage minus the dark voltage.

Next, you'll need to find the sun angle. To do so, hold the case level, with the corner braces up. When you aim the case at the spot on the horizon directly below the sun, the front brace will cast a shadow along your rule. By simple trigonometry, the sun angle is the arctangent (use the `atan()` in your spreadsheet program) of the height of the brace divided by the length of the shadow. Don't forget to also record the time and the sky condition near the sun.

Although you can begin collecting data right away, you will want to calibrate your photometer as soon as possible. Remember that the signal you measure only tells you how much sunlight is reaching your detector. But to measure haze, you need to know how much light isn't making it because it's getting scattered or absorbed. To calculate that, you have to know what signal your instrument would read if you could position it above the Earth's atmosphere. That is, you need to know what signal voltage corresponds to the solar extraterrestrial constant (*ET*). While a bit tedious, it's not hard to find out.

You'll need about half a day, either early morning to solar noon or solar noon to early evening, when the sky is clear blue and there are few or

Aerosol Optical Thickness and the Langley Plot

Aerosol optical thickness (AOT) is the quantity normally used to quantify haze. To calculate it, you must subtract out your photometer's dark voltage and compensate for Rayleigh scattering and sun angle. The equation is:

$$AOT = \frac{1}{M}\left[\text{nl}\left(\frac{ET}{V_s - V_d}\right) - \frac{MP}{8660}\right]$$

where ET is the "extraterrestrial constant" (the signal that would be measured above the atmosphere), V_s is the measured voltage, V_d is the dark voltage, M is the air mass (as calculated from the sun angle), and P is the atmospheric pressure in millibars. The constant 8660 is an empirical factor for Rayleigh scattering at the wavelength detected by the LED.

Langley Plot

no clouds. First, determine the signal voltage and sun angle every 20 minutes when the sun is high overhead and every 10 minutes when it is low on the horizon. Next, calculate the air mass for each measurement. The air mass, *M*, is a measure of the depth of the air column between you and the sun; by convention, the air mass is 1.0 when the sun is straight overhead (sun angle = 90 degrees) and infinite as the sun sets (sun angle = 0). The formula is $M = 1.0 / \sin(\textit{sun angle})$.

To find *ET* for your instrument, simply plot the natural logarithm of each signal voltage against the corresponding air mass. The result is called a Langley plot (see above), and it should be a straight line out to an air mass value of about 10. On this plot, an air mass of 0 corresponds to the top of the Earth's atmosphere. So simply extrapolate the line back to the y-axis. The intercept is the natural logarithm of *ET*. Once you know this number, you can use the equation above to measure something called the aerosol optical thickness (AOT), the critical number scientists use to quantify how much crud there is in the atmosphere. Now it's up to you to publish your data on the web for others to use. Please visit the MAKE site and post a comment at makezine.com/06/citizenscientist if you do.

Dr. Shawn (Shawn Carlson, Ph.D.) is the founder and executive director of the Society for Amateur Scientists (sas.org). He won a MacArthur Fellowship for his work helping ordinary people do extraordinary science.

MAKIN' MAN

Ross Shafer is a busy man. When he's not working in his machine shop, he's working in his woodshop. When he's not woodworking, he's practicing guitar or playing in his band, South Bend (named after a machine in his shop). Or he's riding one of his bicycles (that *he* made), or riding his Ducati through the winding roads of Sonoma County in northern California. Or taking his Ducati apart to replace the lifters. Or refurbishing old amplifiers and old guitars. Or taking metal scraps and making birdbaths, candlesticks, and nightstands for his wife.

"I'm always making something," he says. "I'd rather make something that I need than buy it. If I don't know how to make it, then I learn how."

Right now, he's just finished a steel staircase for a custom house, and is remodeling his own home, a small farmhouse on a gorgeous piece of land that actually *is* a working farm. (There are chickens, sheep, a huge garden with a greenhouse built from salvaged windows, and, at one time, a pig named Homer.) Besides these accoutrements of rural life, there are two refurbished chicken coops and two barns. They contain a music studio, a woodshop, an office, a sewing room, a proper car garage, and a machine shop.

Once the owner of Salsa Cycles, a successful high-end bicycle company known for its hand-built frames and accessories, Ross sold it and began his next career, as a metal fabricator. Or as a rock band guitarist. Or as a sought-after woodworker. Or as an accomplished artist. You can check out this, and his other work, at six-ninedesign.com. —*Shawn Connally*

1. Ross' fixed-gear bicycle (his favorite). He made it. 2. Bedside table he made as a present for his wife, Mari. "It's soooo beauuuuutiful," she says. 3. Tube guitar amps, built by Ross. 4. Side table with attached candelabra. Another gift for his wife. 5. Ross' first self-built tube amp. 6. Big-ass milling machine. 7. The sweet, sweet sound of the first guitar Ross ever built. 8. Ross at one time raced motorbikes; these are his racing leathers, emblazoned with the Salsa Cycles logo. 9. This Ronnie poster was stolen by his wife off a wall in Scotland. 10. Another big-ass milling machine. 11. Ross built this bicycle for his son Max's third Christmas. Max is now 20.

Photograph by Christopher Lucas

1
6
7
8
10

THEORY & PRACTICE

RFID for Makers

Build this kit to read radio frequency ID tags.

By Joe Grand

THEORY

Radio frequency identification (RFID) technologies are rapidly becoming a part of life. RFID tags receive or transmit data (usually a unique serial number) over short distances when scanned by a reader. These tags have worked their way into robotics navigation, inventory tracking, access control, automatic identification, payment systems, and dozens of other applications. Casinos are beginning to embed RFID tags into their gambling chips in order to monitor gambling activity, detect counterfeit chips, and catch cheaters who try to surreptitiously add or remove chips from a wager.

Needless to say, we've seen only the tip of the iceberg for the possibilities of RFID applications, deployments, and projects. The technologies are out there and they're cheap enough for anyone to start experimenting with. Now's your chance to create that access-control-based pet feeder you've always wanted to make. Or maybe you can finally track your kids throughout the house.

In this article, I'll briefly explain the concepts of RFID technologies, introduce the Parallax RFID Reader Module, and show you how to put together a simple, portable RFID "scanner" so you can experiment with RFID and snoop around on your own to discover any hidden or embedded RFID tags.

How RFID Works

(Material in this section is based on information provided by rfidjournal.com.)

There are two major types of RFID tag technologies. "Passive tags" do not contain their own power source or transmitter. When radio waves from the reader reach the chip's antenna, the energy is converted by the antenna into electricity that can power up the microchip in the tag (known as "parasitic power"). The tag can then send back any information stored on the tag by reflecting the electromagnetic waves (think of your access card at work).

"Active tags" have their own power source and transmitter. The power source, usually a battery,

Needless to say, we've seen only the tip of the iceberg for the possibilities of RFID applications, deployments, and projects.

is used to run the microchip's circuitry and to broadcast a signal to a reader. Since active tags have their own transmitter, the reading distance is much longer than with passive tags. Active tags are typically larger and more expensive, and require occasional service. They're commonly used in automatic toll collection systems like FasTrak and E-ZPass. Within both passive and active tag types, there exist read-only, read/write, and cryptographic variations.

Just as you tune your radio to different frequencies in order to hear different radio stations, RFID tags and readers have to be properly tuned in order to

Illustration by Dustin Amery Hostetler/UPSO.org

communicate effectively. RFID systems typically use one of the following ranges: low frequency (or LF, around 125 and 134.2kHz), high frequency (or HF, around 13.56MHz), ultra-high frequency (or UHF, around 868 and 928MHz), or microwave (around 2.45 and 5.8GHz). Generally a higher frequency equates to a faster data transfer rate and longer read ranges, but also means more sensitivity to environmental factors such as liquid and metal that can interfere with radio waves.

There really is no such thing as a "typical" RFID tag. The range of a tag ultimately depends on many factors: the frequency of RFID system operation, the power of the reader, and interference from other RF devices.

RFID for Fun and Profit

There are a few inherent problems with RFID that generally make it not so secure. If you have a reader designed for the correct tag family and frequency, you can communicate with the tag. There is no security between most tag and reader transmissions. This means that you can easily create an RFID scanner, as we'll do in this article, and can snoop around to search for RFID tags (and retrieve the information right off them). If the RFID tags are used for any sort of financial transactions (electronic commerce, stored value), or access control or authentication (storing a unique ID or personal information to prove your identity), you should use them with appropriate caution.

Figure 1. The Parallax RFID Reader Module next to a basic tag.

Photography by Joe Grand

Cracking an RFID Code

In January 2005, the challenge/response scheme of the "secure" Texas Instruments Digital Signature Transponder (DST) tag was cracked. The proprietary cryptographic cipher used in the challenge/response scheme was successfully reverse engineered by some folks at RSA Laboratories and Johns Hopkins University based on a single PowerPoint slide inadvertently discovered on the web. The tag is used around the world in such applications as vehicle immobilizers and the Mobil Speedpass gas pay system. Details of the attack can be found on the "Analysis of the Texas Instruments DST RFID" website at rfidanalysis.org.

Not only did these researchers crack open the secret egg with their knowledge of the algorithm and a distributed, brute-force key search using 100MHz field-programmable gate arrays (FPGAs), but they also created a system to allow them to eavesdrop on the communications between a DST transponder and reader, or to participate actively in the protocol by emulating either device. What exactly does that mean?

Over 150 million deployed DST devices are now at risk and could be cloned or spoofed. Free gas for everyone!

Mark of the Beast

Meet Amal Graafstra, a maker who embedded an RFID tag into his hand. Now, Amal can unlock his house door without reaching for his keys and log into his Windows box without having to remember a password. (His interview with MAKE can be found at makezine.com/go/amal.)

The "Tagged" RFID implant forum (tagged.kaos.gen.nz) has been created to allow folks to share information and their experiences regarding implanting RFID technologies into their own bodies.

Certainly not for the faint of heart, this is a far cry from the civil liberties and personal privacy communities who consistently push back against legislation enabling embedded RFID tags in such personal items as cash and passports. What better way to have your privacy invaded than by doing it yourself?

PRACTICE

This will give us a portable RFID scanner that you can use as a "bug finder" to snoop through your house and scan yourself and your friends, family, and pets to check for RFID tags.

The Parallax RFID Reader Module

I designed the Parallax RFID Reader Module (Figure 1, previous page) as a low-cost solution to read passive, low frequency (125kHz) RFID tags from up to 3 inches away. The module requires a single +5VDC supply and it interfaces to a PC, BASIC Stamp, Microchip PIC, or other processor via a 1-wire, 2400 baud serial TTL interface. It costs $39 from Parallax (parallax.com, part #28140) and can be integrated into any design using only four connections (VCC, /ENABLE, SOUT, GND). It just can't get much easier.

An LED shows the state of the reader module. When the module is successfully powered up and is in an idle state, the LED will be green. When the module is in an active state and the antenna is transmitting, the LED will be red.

The RFID Reader Module is controlled with a single TTL-level active-low /ENABLE pin. When the /ENABLE pin is pulled low, the module will enter its active state and enable the antenna to interrogate for tags. Keep in mind that when the antenna is active, the device is broadcasting and consuming about 100 to 200mA from the power supply. If you're going to build a project that involves batteries, you may want to add a physical button to activate the reader only when a card is actually present, or use a timeout in your controlling software/firmware to disable the reader periodically and reduce the load on the power supply.

This particular product is designed to detect common, passive read-only tags within the EM Microelectronics-Marin SA EM4100 family. When a valid RFID tag is placed within range of the reader, the tag's unique ID will be transmitted as a 12-byte ASCII string via the TTL-level SOUT (Serial Output) pin in the format shown in Table 1.

The start byte and stop byte are used to easily identify that a correct string has been received from the reader (they correspond to a line feed and carriage return characters, respectively). The middle ten bytes are the actual tag's unique ID.

All serial communication is non-inverted at 8 data bits, no parity, 1 stop bit (8N1). The baud rate is configured for 2400bps, a standard communications speed supported by almost any microprocessor or PC, and cannot be changed. The RFID Reader Module initiates all communication and can connect directly to any TTL-compatible UART.

If you plan to connect the RFID Reader Module directly to a PC's serial port, you'll need a common RS232 line driver (also known as a "level shifter"), such as the Maxim MAX232 or Linear Technology LT1780, to bring the module's TTL-level outputs to the PC's expected range (Figure 2).

Making Something Fun

Armed with our knowledge of RFID and the Parallax RFID Reader Module, let's make something. Because the RFID Reader Module simply sends a 12-byte ASCII string containing the tag's unique ID, we can connect the output of the reader to the input of a serial LCD module, such as Parallax's 2x16 Serial LCD regular (part #27976) or backlit model (part #27977), which will display whatever ASCII characters it receives from its input pin onto the LCD.

Combining the RFID Reader Module and the serial LCD will give us an instant, portable RFID scanner that you can use as a "bug finder" to snoop through your house and scan yourself and your friends, family, and pets to check for RFID tags. The best thing about this project is that no software is required! The RFID Reader Module itself handles all of the underlying RFID-reading functionality, and the serial LCD is already expecting to receive serial data.

To build our project, simply hook up the two pieces as shown in Figure 3. If you're using a 9V battery or any other power source above 5V, you'll need to use a voltage regulator (like the standard 7805) to bring the voltage down to 5V. Don't even think about connecting an unregulated power source directly to the RFID Reader Module or LCD!

Your final product will look something like Figure 4. I built mine on a simple breadboard to show off at the Make Fest at O'Reilly's Emerging Technology Conference in San Diego (March 2005) and still haven't taken it apart. The text string shown on the LCD, "04129C1A89," corresponds to the unique ID of the most recent tag held up to the reader.

With your RFID scanner ready to go, you can begin your hunt for RFID tags — hidden or otherwise.

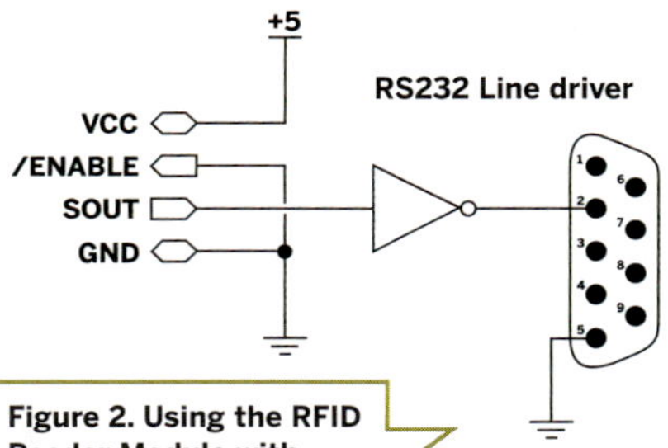

Figure 2. Using the RFID Reader Module with a PC.

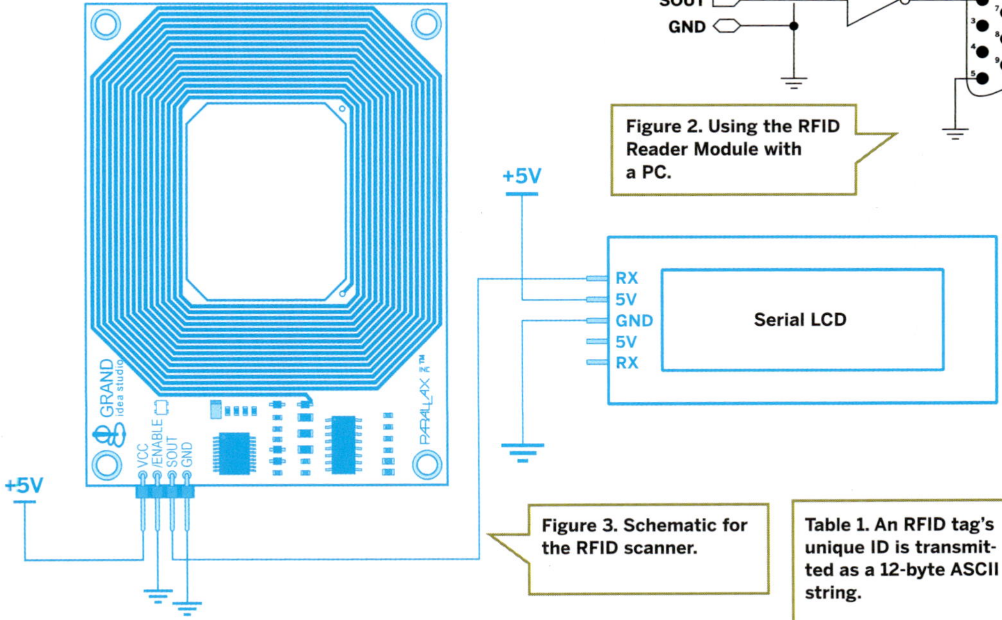

Figure 3. Schematic for the RFID scanner.

Table 1. An RFID tag's unique ID is transmitted as a 12-byte ASCII string.

MSB											LSB
Start Byte (0x0A)	Unique ID Digit 1	Unique ID Digit 2	Unique ID Digit 3	Unique ID Digit 4	Unique ID Digit 5	Unique ID Digit 6	Unique ID Digit 7	Unique ID Digit 8	Unique ID Digit 9	Unique ID Digit 10	Stop Byte (0x0D)

Simply walk around to find any tags within range, or modify the design to work with your specific project or application.

NOTE: Spurious "reads" can be caused by RF interference from nearby electronic devices. Also, text on the LCD can eventually get jumbled, due to the way the LCD handles the start and stop bytes — if so, just reapply power.

For information on interfacing the RFID Reader Module with a Parallax BASIC Stamp to make some interesting access control projects, check out Jon Williams' article, "You Can't Touch That: Non-Contact Access Control," in the April 2005 issue of *Nuts & Volts* magazine.

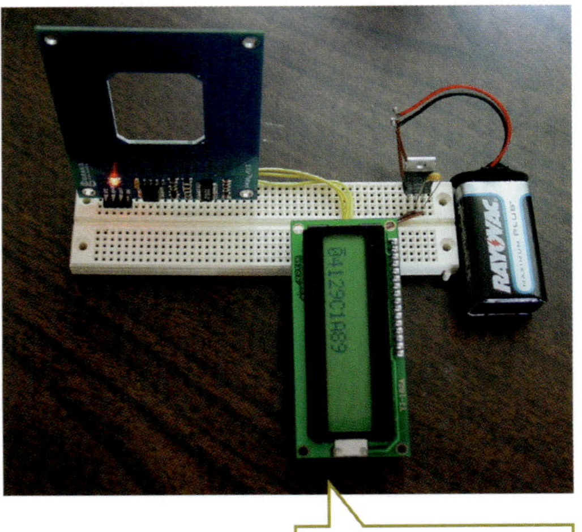

Figure 4. RFID reader module in action.

Joe Grand (joe@grandideastudio.com) is the president and principal electrical engineer of Grand Idea Studio, Inc., a product research, development, and licensing firm. He specializes in the invention and creation of consumer electronics, video game accessories, and toys.

Illustrations by Joe Grand and courtesy of Parallax, Inc.

MAKING

XM58AB
LM3940IS
-3.3 P+
M9730AJ
LMC6484AIN
M9730AJ
LMC6484AIN
AT91SAM7X256
AU-ES
ARM
MAKE
V0.9
ON
1 2 3 4 5 6 7 8
CTS 208-8 T542

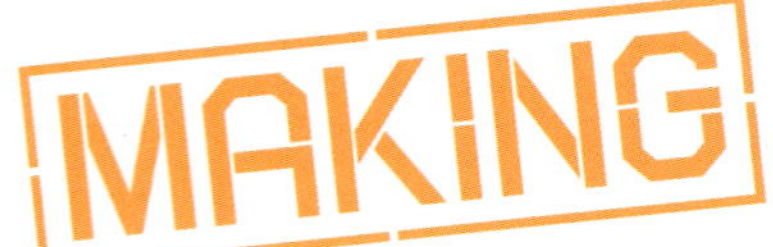

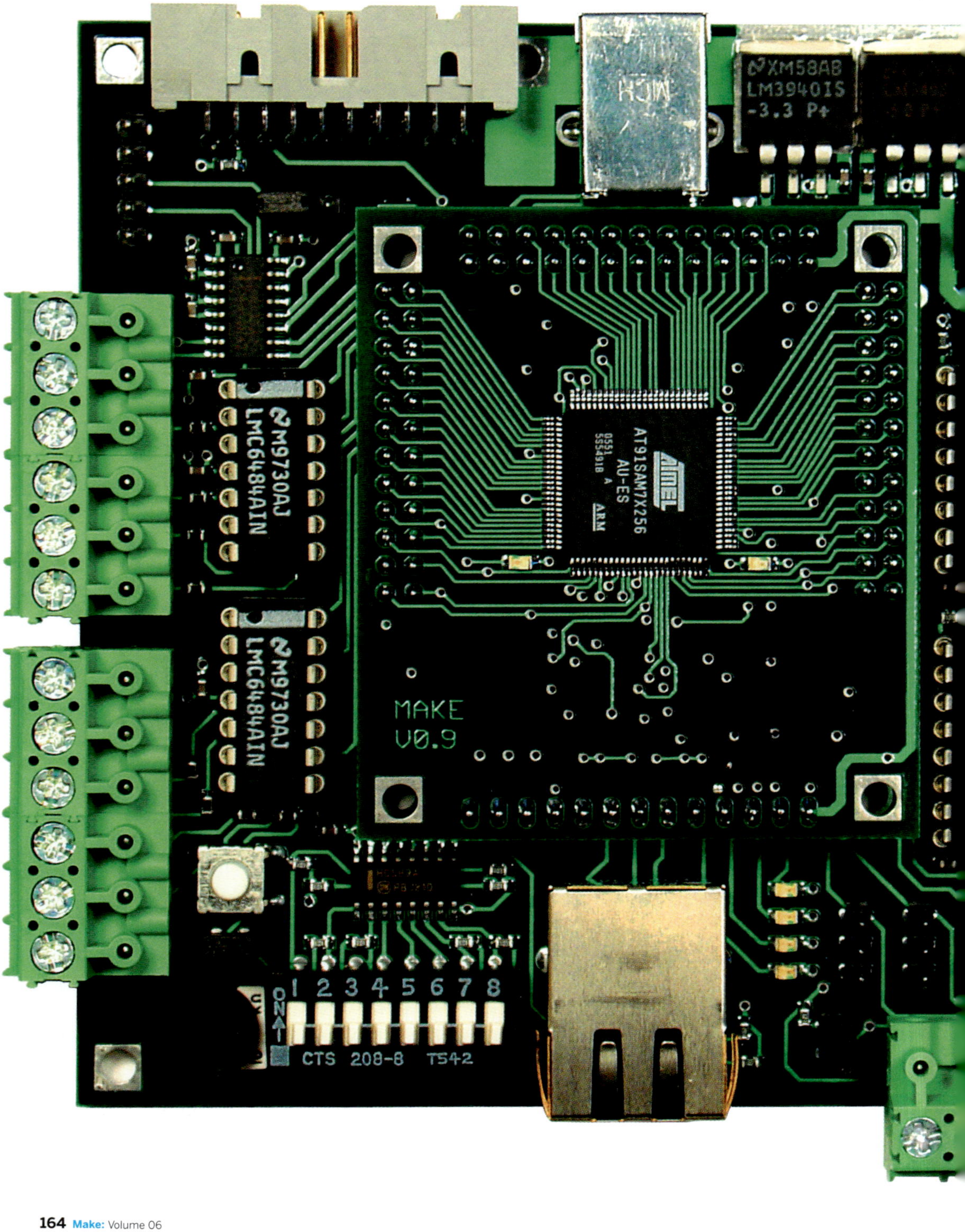

The MAKE Controller

Announcing a just-maybe-revolutionary microcontroller for all things DIY.

By David Williams and Liam Staskawicz

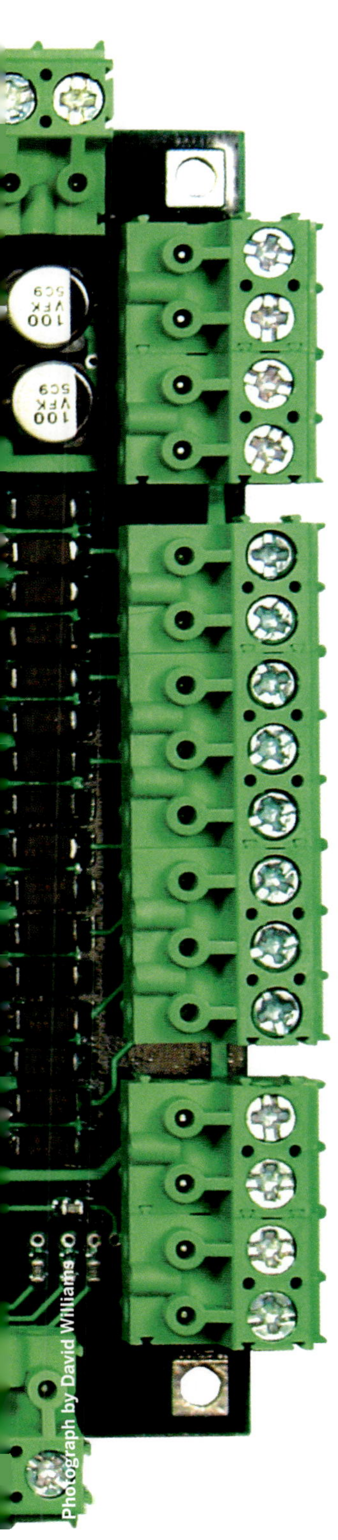
Photograph by David Williams

Microcontroller chips and the tools required to use them were once so specialized that few people could explore their potential without a degree in electronics and software engineering.

From BASIC Stamp kits to Wiring and Arduino boards, manufacturers and open source communities now offer somewhat easier ways to program microcontrollers, enabling cross-pollination between the engineering world and hobbyists, do-it-yourselfers, artists, and students.

But the effort needed to program a chip and design a circuit is still a barrier to countless creative people, even as the latest 32-bit chip designs offer new features like built-in networking.

All the elements are in place for the next revolution in DIY microcontrollers: a friendly, open platform that can do it all, launch a thousand projects, and empower both novices and pros.

Open Source Hardware

It's always exciting when powerful technologies become accessible to ordinary people. This is happening now with microcontrollers. After revolutionizing many industries (and requiring industrial-sized development budgets), now these do-anything chips are becoming easier to program and use with free, open source tools.

We realized, as did MAKE, that a general-purpose controller kit served by a free and open development environment would offer limitless potential to makers. A smart building block like this would shortcut (and even enable) diverse projects, collapsing pages of complicated instructions into, "connect these components to the controller board, upload this code, and *voilá* — you've got an amazing new device for your kitchen" (or garden, or living room, or backyard, or ... *see sidebar, page 169*). This platform would also inspire more advanced makers to develop and share their own original projects. Ideally, such a controller kit would be visible enough to attract a healthy developer community, and rewarding enough to see it grow.

Last year, MAKE magazine approached MakingThings to create this platform, the MAKE Controller. We were delighted, and we're even more delighted to announce that it is now available. In this article, we explain how the kit's two PC boards work, and why we designed them the way we did.

Basic Principles

Right from the outset, it was imperative that this be an open project. We knew from our experience with similar products, and from our customers (MakingThings has been producing and selling a range of controller boards since 1998), that anything short of freely available hardware and software would frustrate the creativity of the experimenter and the hacker — precisely the audience we're aiming for. We will publish all of our schematics, to encourage modifications and improvements to the hardware. The software environment(s) had to be free and easily accessible as well, to promote development. By supporting all aspects of the project with cross-platform, free, and open tools, we created an open hardware platform that many disparate communities can support, contribute to, and enjoy. For further accessibility, we also wanted the kit to be relatively inexpensive, and we managed to keep it down to less than $150.

To serve both novice users and experienced engineers, we decided to base our design around two boards rather than one. The Controller Board contains the microcontroller itself, plus the delicate circuitry needed to run it it, and all essential peripherals. This Controller Board can then plug into a larger Application Board that provides circuitry to interface with real-world electronics like sensors and actuators. Users of an Application Board can just plug a stepper motor or a pneumatic valve right in, rather than having to design the circuitry to drive them from the chip. We designed one general-purpose Application Board with a nice mix of control and communication capabilities, and over time we will add others with different features, such as CMOS cameras and audio input and output.

> We decided on the newer, faster class of chip, to create a new generation of controller boards.

With the two-board approach, novices can experiment with an Application Board with little fear of breaking anything, while experts can either use an Application Board or plug the Controller Board into a board of their own design.

For the most advanced programmers, the Controller Board makes it easier to get at the microcontroller's densely packed pins. Everything is accessible. They can read the 700-page processor manual, dig deep into the chip's architecture, choose or write their own software and device drivers, and have full control over every other aspect of the board's operation.

Novices also have options. The Application Board includes onboard switches that select and control useful and fun software that's pre-installed on the Controller Board. This makes the kit enjoyable and configurable right out of the box, without any programming. After exhausting the switches, a novice's next step can be downloading different code and uploading it into the boards using USB.

Liam and David with the final MAKE Controller Kit and several prototypes.

Choosing the Microcontroller

The first and most difficult task was choosing the microcontroller. Most microcontrollers that address the hobby/artist/DIY markets are 8-bit systems, in which most operations in the processor work on 8 or 16 bits at a time. All of the predecessors of the MAKE Controller are 8-bit, including our own products and the Wiring and Arduino boards.

32-bit microcontrollers have recently become available, some of which are ten times faster than 8-bit chips, with far more onboard memory and sophisticated on-chip resources. They can comfortably multitask and communicate over bridges and networks, in addition to running simple control operations. On the downside, they're also a bit more expensive (about $9 rather than $7 or less) and they use more power.

We decided on the newer, faster class of chip, to create a new generation of controller boards. Then we had to choose a specific one, since Philips, STMicroelectronics, Texas Instruments, and Atmel all have excellent 32-bit products. Several meetings and a masochistic one-day trip from San Francisco to Boston later, we had enough information to select our chip: the Atmel AT91SAM7X256 — or "SAM7X." This is a brand-new part with a staggering list of features that hobbyists could only dream of before its release. In addition to quite exciting basic stats (a 32-bit 55MHz ARM core, 256KB of flash memory, and 32KB of RAM), the part has Ethernet addressability, a USB port, a CAN controller (CAN = Controller Area Network, a simple, flexible, two-wire protocol originally developed for cars), several serial ports, and more, making it one of the most versatile microcontrollers on the market.

The SAM7X's Ethernet port opens up entire realms of possible applications — for example, it can host a web server or telnet session, or retrieve data from the internet all by itself. And its USB port lets the system interface with a computer, to offload any computational heavy lifting. With either port, you can have a computer perform specialized processing while the connected microcontroller exchanges information with the rest of the world.

Finally, because of early excitement around the introduction of the SAM7X, many software providers have adapted their development environments and other tools to this chip, resulting in strong software support even before we got started.

Selecting the Development Environment

The next thing we did was confirm that there was a full set of free tools that would allow people to develop with the SAM7X. Fortunately, GNU's incredible open source compiler, gcc, will compile code for most ARM processors. With every release, gcc becomes more capable and more efficient, although at the time of writing, the debugging environment that gcc provides for ARM parts is somewhat limited.

A few expensive professional tools also target the SAM7X processor and use gcc, including Rowley and Associates' CrossWorks for ARM. This means you can use either a high-end environment or open source tools to develop for the MAKE Controller.

Designing the Hardware

By far, the hardest part of the process is designing the hardware. Here are the steps we followed:

1. **Decide on basic functionality.** What will each board include? One such decision was to squeeze all Ethernet support (except for the connector itself) onto the Controller Board, making it available to all users, instead of putting it on the Application Board.

2. **Decide which microcontroller pins do what.** The SAM7X has 100 pins, but that's not much, considering what the chip can do. Manufacturers rely on

Photograph by Susan Williams

MAKE CONTROLLER KIT: CONTROLLER AND APPLICATION BOARD LAYOUTS

Layout of Controller Board plugged into Application Board. **Dashed connectors are detachable.**

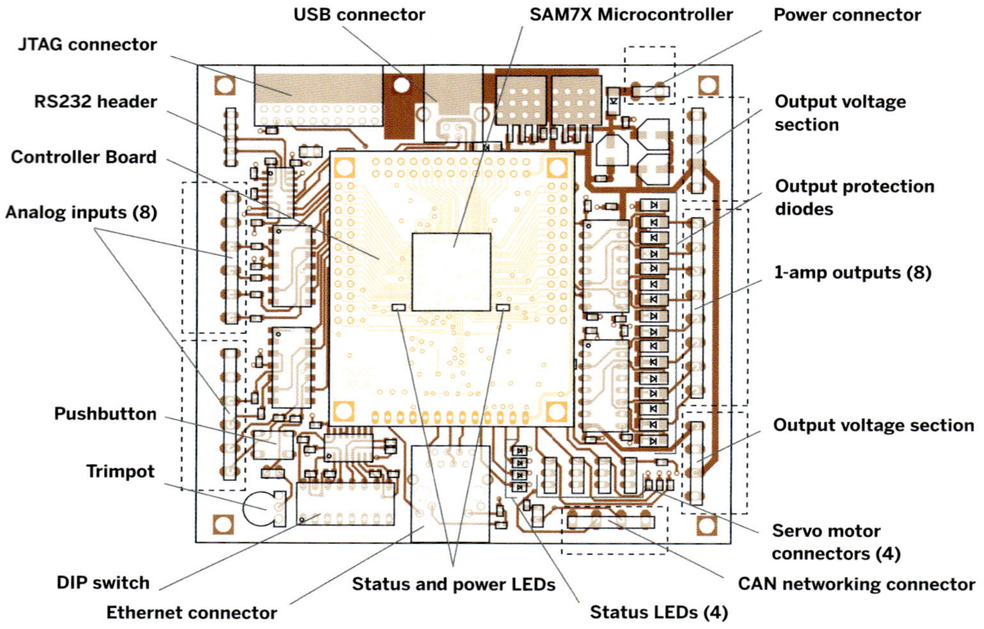

pages worth of usage modes and dependencies in order to squeeze more functionality onto their chips, so the task of allocating all the signals becomes a monumental logic puzzle.

3. **Work out power issues.** What power will be needed by what parts? What is the maximum voltage the board should withstand?

4. **Draw schematic.** Wire all components per manufacturer's notes and common design practices.

5. **Select packages for all devices.** We favored surface mount (SMT) components since they're smaller, cheaper, and cheaper to assemble — although you can't solder them easily by hand.

6. **Map schematic onto a PCB board.** Each trace connects two points, and they can't overlap or run too close. Ground and power can go on the internal layers of a four-layer board, as a wiring convenience and to reduce interference from electrical noise.

7. **Check and double-check everything.**

8. **Generate Gerber files.** PCB fabricating machines use these to manufacture boards. Most PCB layout software has a Gerber export option.

9. **Send Gerber files to a board shop that can do the job.** The traces on the Controller Board, and even the pads of the SAM7X itself, are very close together. With a design this fine, many cheaper shops will send boards back with shorts between contacts.

10. **Assemble the boards.** Order all the parts, and send them with the boards to a manufacturing facility.

11. **Test.** With our first boards, we found mistakes in the complex Ethernet area. Correcting these required some real white-knuckle soldering tricks.

12. **Celebrate!** Designing a new board is tense, and failure means a loss of money as well as pride. When it all finally works, everyone breathes a sigh of relief.

Illustration by Liam Staskawicz

Designing the Software Environment

One of the difficulties of programming a microcontroller directly is having to write for the device's specific inner workings. This means structuring code to permit other functions to operate, and deciphering arcane header files and the bit-by-bit operation of complex peripherals. We wanted none of that.

To liberate our users (and ourselves), we wrote a series of code layers that access all the devices on the boards at a more abstract level. Instead of needing to understand every aspect of each device, developers can include these files with their projects, and then call simple methods that encapsulate lots of fiddly hardware settings. For example, with one simple call, our `AdcInit()` routine initializes the SAM7X's Analog to Digital Converter, a process which otherwise requires a hideous chunk of code.

Another difficulty with bare-bones microcontroller programming is having to explicitly manage all control-passing, sharing the processor among all running tasks. The cure for this pain is to use a Real Time Operating System (RTOS). With complex projects where functions run simultaneously or timing needs to be managed closely, an RTOS makes coding applications quicker, simpler, and more elegant.

RTOSs have traditionally been expensive, and they consume too much overhead to be practical with 8-bit microprocessors, but there is an excellent, free RTOS, helpfully called FreeRTOS.org, that runs on the SAM7X. When connected to lwIP, a small library of code designed to provide internet capabilities, we have a full RTOS with internet capability.

Programming with the benefit of an RTOS means you can start tasks upon initialization, and run them concurrently as permanent loops that poll for input and communicate with each other. There's no need for spaghetti code that breaks out of each ongoing task to check on everything else.

Conclusion and Future

We are very pleased with the MAKE Controller Kit. It is an absolute delight to program, and connecting real devices to it is very simple. We look forward to adding more and more functionality, and seeing what others think of doing with it. Try one and tell us! Email us at info@makingthings.com.

Project Possibilities

Some applications of the MAKE Controller that you may be seeing in the near future.

Drink-O-Mat
Timed control of valves and motorized swizzle stick can dispense and mix measured amounts of different beverages from inverted bottles.

Plant Cyborg
Light sensor and motor controllers position houseplant for optimal growth, while moisture sensor triggers valve to ensure perfect watering.

Glue Gun 3D Imaging
3-axis stepper motors and trigger controller render stable 3D forms by building them up layer-by-layer using hot glue.

Automated Pet Feeder
Motor device feeds pet remotely via web-based interface or timer program, and dispenses treats when pet stands in a specific position.

Liquid Temperature Controller
Thermostat feedback loop keeps liquid within a specified temperature range — for brewing, aquariums, candy-making, photo developing, etc.

Global Sensor Network
Large-scale survey system registers sensor input from hundreds of physical nodes connected to a common server, which publishes sensor info.

Get Your MAKE Controller Kit!
On sale now at makezine.com/controller, where you can also learn more about the kit, including full schematics, and share your experiences in our online forum.

David Williams enjoys the connection between machines and intelligent software, and has designed and built professional tools for software, electronic, and mechanical engineering. Liam Staskawicz has explored sensor-driven hardware and software systems from the perspectives of both engineer and musician. See makingthings.com.

BY SAUL GRIFFITH, NICK DRAGOTTA & JOOST BONSEN '06

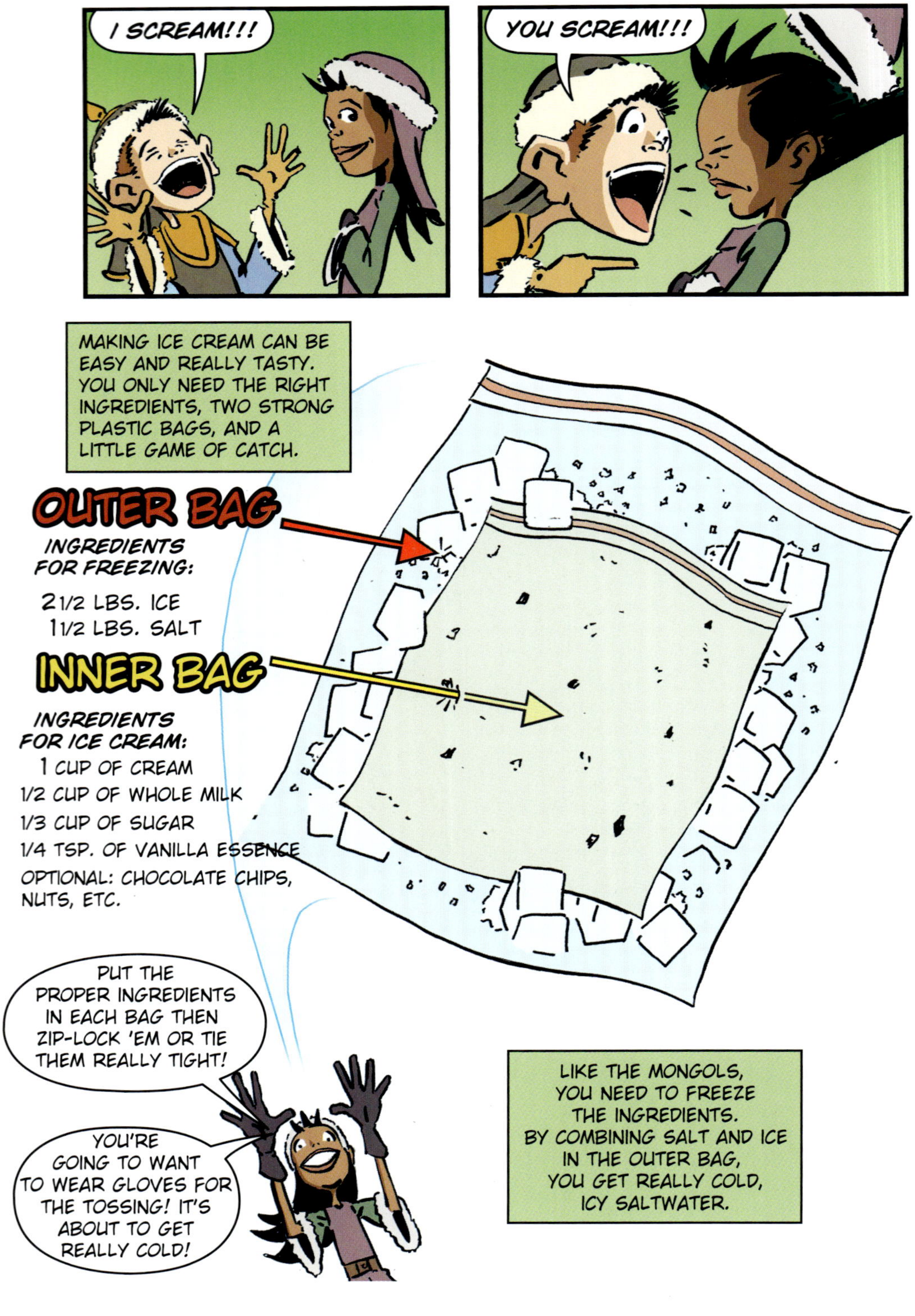

MAKING ICE CREAM CAN BE EASY AND REALLY TASTY. YOU ONLY NEED THE RIGHT INGREDIENTS, TWO STRONG PLASTIC BAGS, AND A LITTLE GAME OF CATCH.

OUTER BAG

INGREDIENTS FOR FREEZING:

2 1/2 LBS. ICE
1 1/2 LBS. SALT

INNER BAG

INGREDIENTS FOR ICE CREAM:

1 CUP OF CREAM
1/2 CUP OF WHOLE MILK
1/3 CUP OF SUGAR
1/4 TSP. OF VANILLA ESSENCE
OPTIONAL: CHOCOLATE CHIPS, NUTS, ETC.

PUT THE PROPER INGREDIENTS IN EACH BAG THEN ZIP-LOCK 'EM OR TIE THEM REALLY TIGHT!

YOU'RE GOING TO WANT TO WEAR GLOVES FOR THE TOSSING! IT'S ABOUT TO GET REALLY COLD!

LIKE THE MONGOLS, YOU NEED TO FREEZE THE INGREDIENTS. BY COMBINING SALT AND ICE IN THE OUTER BAG, YOU GET REALLY COLD, ICY SALTWATER.

From the forthcoming book *HOWTOONS Book 1* by Saul Griffith, Nick Dragotta, and Joost Bonsen. Published by arrangement with ReganBooks, an imprint of HarperCollins Publishers, Inc.

MakeShift

By Lee D. Zlotoff

The Scenario: You live with your wife and two teenage children in a rustic, one-story house next to a roaring mountain stream in a steep, narrow canyon. Early one morning, you're awakened by a serious earthquake. The family is OK, but the power is out. The portable radio tells you that the quake has hit the region hard, and power will probably be out for days. The land and cellular phone systems are out as well, but you don't get cell reception up here anyway.

You look outside and see that a rockslide just below you has not only cut off your driveway, but also dammed the stream, which is rising fast. Your house will be mostly underwater in 15 minutes.

You have two cars: A Toyota Prius and a Chevy Suburban. Both have full tanks of gas and cellphone chargers in the dash, but you have only enough open space up behind the house to keep one of them out of the water, above the level of the rockslide. Your home has typical belongings and a basic set of tools, but all your camping equipment is lent out to your brother — and it's been getting cold at night.

The Challenge: You have 15 minutes to move one of your cars up to dry ground, and tell your family what to collect from the house to ensure that you can all survive up here for at least five days. You also need to devise a way to let the outside world know that you're here, and need to be rescued.

There's one other wrinkle: Your only neighbor, a single guy named Dave, probably partied hard last night and slept through the quake. But his house will flood just as quickly as yours. You can send someone to rouse Dave, but then you're down one collector, and if you do rescue him, he'll join you with nothing more than the clothes he has on. Should you factor in Dave, or figure that in a disaster like this, it's every man for himself?

Now, look at your watch again. You need to make some key decisions immediately, and figure out who should do and grab what. Because in 15 minutes, the party's over, and you're completely isolated and homeless ... Go!

Send a detailed description of your MakeShift solution with sketches and/or photos to *makeshift@makezine.com* **by July 17, 2006. If duplicate solutions are submitted, the winner will be determined by the quality of the explanation and presentation. The most plausible and most creative solutions will each win MAKE sweatshirts. Think positive and include your shirt size and contact information with your description. Good luck! For readers' solutions to previous MakeShift challenges, visit** *makezine.com/makeshift*.

Lee D. Zlotoff is a writer/producer/director among whose numerous credits is creator of *MacGyver*. He is also president of Custom Image Concepts.

Photograph by Topher Lucas

GASOLINE

The best tools, software, gadgets, books, magazines, and websites.

Turn your living room into a high-tech pool of primordial ooze.

Roboraptor Rules the Roost

$100 roboraptoronline.com

Guided by the design brilliance of Mark W. Tilden, WowWee Robotics created the best-selling robot of all time, Robosapien. Dubbed the only robot fluent in "caveman talk," Robosapien has now "evolved" into the dinosaur-like Roboraptor.

Bristling with IR, touch, and sound sensors, the 32-inch-long Roboraptor is fully equipped to interact with its environment. A remote controller, fondly reminiscent of a PlayStation 2 game controller, controls body movements, personality modes, and dino moods. Forget the Prozac: mood swings are encouraged with Roboraptor. Hunting, playful, and cautious moods are simple two-button commands. Unlike Robosapien, Roboraptor has a special "laser" targeting feature that can be accessed with the remote controller. Just point the green "targeting assist light" on any surface, like a wall, and Roboraptor will hunt down the target.

Is Roboraptor as hackable as Robosapien? I asked Mark Tilden, and discovered that you can saddle up Roboraptor with a Barbie or a G.I. Joe, add a microcontroller for responding to the dino's impressive sensor array (a higher brain function hack), and design a Ni-Cd wall charger for Roboraptor to stalk. Now get out there and make it.

—Dave Prochnow

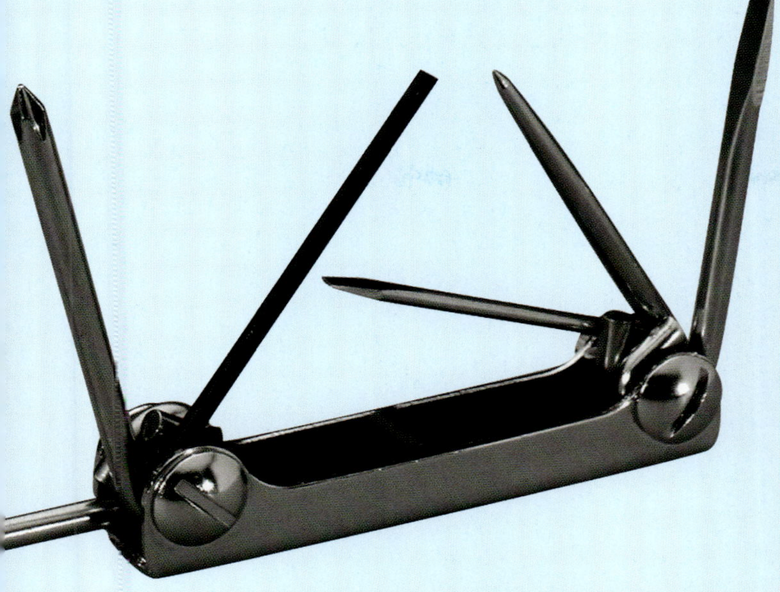

The Übergeek Wondertool

Eklind 6-Blade Screwdriver $12 eklindtool.com

While it looks like something you would put in your glove box and forget, I've kept my original Eklind 6-blade Screwdriver in my pants pocket for more than 15 years and have used it almost every day.

I've adjusted headlights on a dark road with the T-15 Torx, cracked thousands of computer cases with the #1 Phillips, fixed eyeglasses with the small slot, and mounted routers using the awl to drill starter holes in mounting boards. While it doesn't have pliers like some wannabe geek tools, it is far superior in terms of weight, choice of blades, and feel in your hand. *—Steve Johnson*

Maker Heaven

Ax-Man Surplus Stores ax-man.com

I moved from Minneapolis a few years ago and I miss this store so much. They resell industrial parts, wheels, motors, speakers, fabric, foam, biohazard suits, and just about anything you can use to make stuff. As you walk through Ax-Man, you'll notice large motorized moving displays with baby doll parts and humorous descriptions accompanying their products. I bought some beautiful, large Hollywood-style metal film canisters there and sanded them down to make a portfolio case. Twin City residents, take note.

—Joel Raedeke

Government Leftovers

govliquidation.com

Looking for "Space Survival Equipment" or "Nuclear Reactors"? Government Liquidation is the place to go. Just be prepared for a bit of paperwork if you bid on anything from the "Combat Ships & Landing Vessels" category.

The selection is wide: lathes, presses, steam turbines, saws, tractors, cranes, tents, boats, jeeps, bulk computers, solenoids, amplifiers, sirens, motors, teletypes, cameras, oscilloscopes, gauges ... in short, everything a geek could dream of.

Beware of misleading picture sizes. I once purchased a Hughes FACT PC (a giant DEC PDP-11 system used for testing aircraft circuitry) that included a box of cables. The box turned out to be 6 feet square and 4 feet high. Some items are labeled "bring your own crane." Also, the government doesn't ship, so make certain the item you're bidding on isn't in Guam. Most of the equipment is stored at military bases. Be sure you bring ID and be prepared to have your vehicle searched.

It can be a hassle to deal with, but Government Liquidation is an incredible source. It's the sort of site that makes you wonder, "What would I do with 300 snowshoes, a Gould Instruments Thermal Array Recorder, and a 1970 AM General Cargo Truck?"

—Tom Owad

MAKE LOOKS AT KITS

Editor Phillip Torrone talks about kits that caught his eye on makezine.com.

9V USB Charger Kit

$9 aarondunlap.com/blog/1130885615

Aaron Dunlap made a great little kit for folks who'd like to make their own 9V USB charger. And, you guessed it, it's all inside an Altoids tin. It's pretty simple to construct one of these, but Aaron put everything together in kit form if you're not a hardcore maker yet.

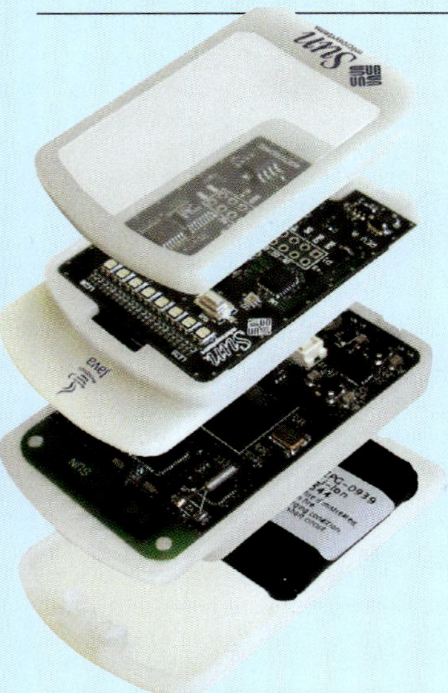

Sun SPOT Kit

$499 sunspotworld.com

This Java-based wireless-sensor development kit, an ongoing project from Sun Research Labs, is all about sensors. It's pretty techie, but looks like a lot of fun. Release is scheduled for May 2006, and people are using the Sun SPOT forum already.

Build Your Own Digi-Comp!

$55 mindsontoys.com

Tim Walker missed the classic 3-bit Digi-Comp mechanical computer from the 1960s, so he figured out how to make his own. Now he's reissued it as a kit.

Wooden Clock Kits

wooden-gear-clocks.com **and** krazydad.com/blog/2006/01/24/

One of our readers wrote in and suggested a few wooden clock websites for kit builders. He started out with the Ascent, which is precut and ready to build. (Plans are also available for those who want to get down and dirty with their jigsaws.) The finished product looks gorgeous.

Carnivorous Plant Kit

$25 scientificsonline.com

Edmund Scientific has a deal on Audrey II meat-eating plants. Their kit lets you create an "authentic" bog for growing various carnivorous plants from seed in a specially designed terrarium. Feed them, Seymour!

Arty Electronics for Kids

$250 playfulinvention.com

I saw an early demo of these tiny computers at the Exploratorium in San Francisco, and they're amazing. The PicoCricket Kit is designed for fusing art and robotics using lights, sound, music, and motion.

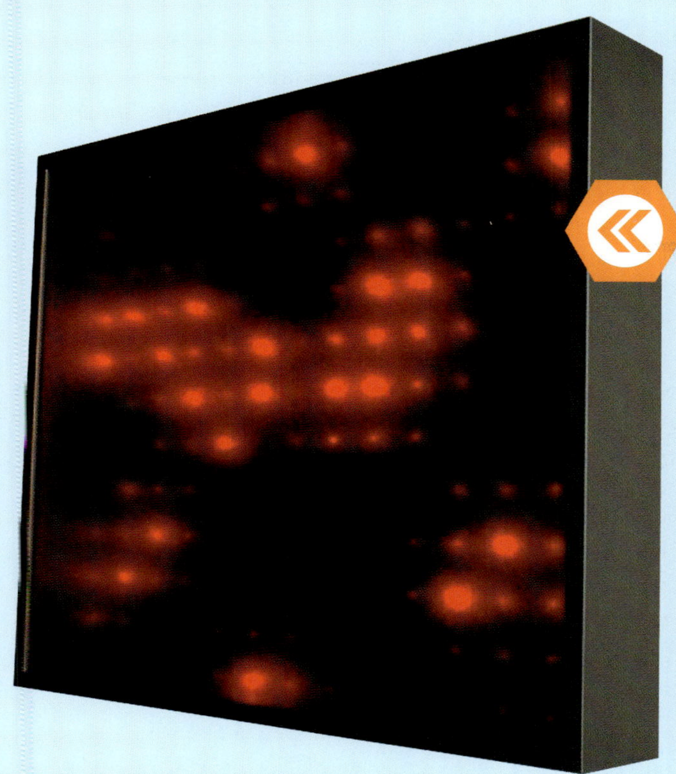

DIY LED Panel Kits for Artists

$235–$359 biteditions.com/diy.html

Send off your own animations to Bit Editions, and then assemble the kit to display your LED-ified art. The displays are 12x14 inches, and contain one LED per square inch, with each LED capable of eight levels of brightness. Driven by a microcontroller and 128K of programmable memory, the panel gives artists up to 1,984 frames for animation. The new version, scheduled to be on sale in May, will be reprogrammable from a PC through a serial port.

Lie Detector Kit

$11 apogeekits.com/lie_detector.htm

We've had a few emails and posts from folks who wanted to build their own lie detector, and now you can. The website explains how lie detectors work (by measuring the change in skin resistance of the subject — you sweat when you're lying) and also has a circuit diagram.

Blow Your Mind

Lego Mindstorms: Robotics Invention System 2.0 $199 lego.com
Contrary to what you might have thought, you can do a lot more with Legos than build another useless static display. You can pick up a set that includes a bunch of gears, motors, sensors, and a fairly sophisticated microcontroller that runs all of this stuff for under a couple of Bennies. But don't just settle for what comes out of the box. If you really want to impress the girls, be sure to reprogram the firmware with an alternative like LeJOS to really squeeze out that last bit of juice. Also, keep an eye out for some fat price cuts coming up, since Lego announced a newer version of this set coming out later in the year.

—Matthew Russell

TOOLBOX FAVORITE THINGS

Chosen by the editors of *Ambidextrous* magazine.

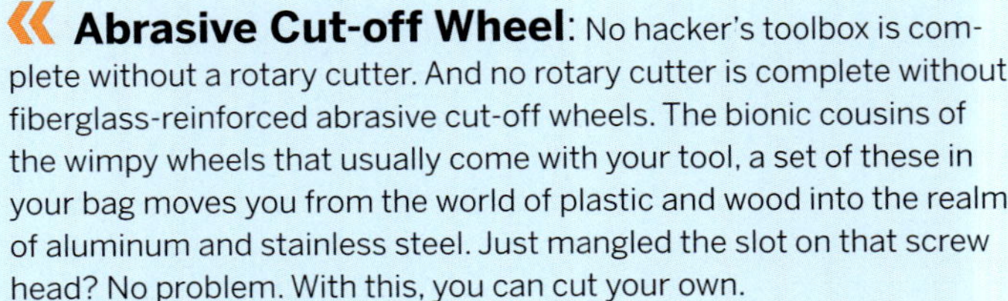

Abrasive Cut-off Wheel: No hacker's toolbox is complete without a rotary cutter. And no rotary cutter is complete without fiberglass-reinforced abrasive cut-off wheels. The bionic cousins of the wimpy wheels that usually come with your tool, a set of these in your bag moves you from the world of plastic and wood into the realm of aluminum and stainless steel. Just mangled the slot on that screw head? No problem. With this, you can cut your own.

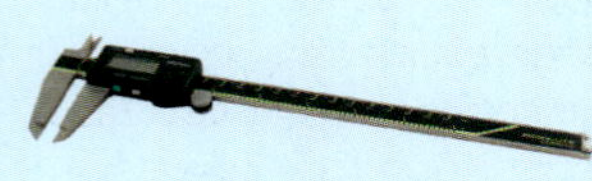

Digital Calipers: Whether you're gauging the width of gaps, the diameter of pegs, or the depth of holes, calipers are immeasurably more useful than rulers. It's true that the digital variety require batteries, but the ease of use and accuracy of digital calipers will save you a lot of heartache when you've been working too late, too long, and too hard on a project.

Gaffer's Tape: Move over, duct tape — gaffer's tape is as versatile as its Space-Age cousin, but looks nicer and doesn't leave behind a sticky, messy residue when it's pulled up. Absolutely indispensable when you're attaching your inventions to something valuable, like a brass instrument or someone else's carpet. Gaffer's tape also rips easily but secures firmly. The catch? It's pretty pricey, about $20 a roll.

Spray Can Trigger Handle: No more tired index fingers! With the Spray Can Trigger Handle, avoid the frustration of keeping even pressure applied to the plastic nub on the top of the spray paint can. This handle transforms any cheap can of paint into a much-easier-to-use, airbrush-like system. Your projects can now be professionally primed and painted. And at less than $3, the Spray Can Trigger Handle is a cheap must-have for any paint shed or toolbox.

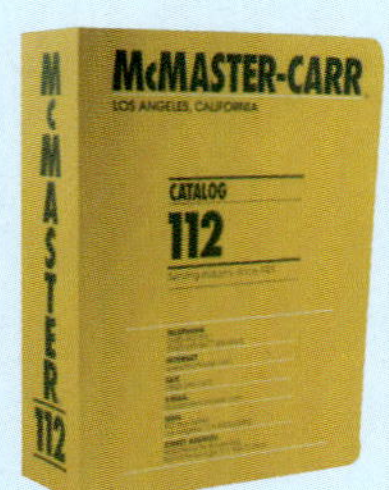

McMaster-Carr Catalog: In our book (a big yellow and green one, in fact), McMaster-Carr trumps all other industrial supply companies. You can get anything from McMaster. Standard shipping is surprisingly speedy, with orders usually magically appearing on our doorstep in a day or two. And, there is no better source of brainstorming fodder than their well-organized catalog chock full of extra-descriptive info and masterful line drawings. Can't get your hands on one of their coveted catalogs? Their website is excellent as well: mcmaster.com.

Safety Glasses: Actually, safety glasses are more of a favorite thing by association. What we are truly in love with is our eyes. They've been invaluable in most every project we've done to date, and given their all-around utility and the fact that they're amazingly hard to replace, we're committed to protecting the ones we've got.

When not distracted by their advisers' urgings to finish their Ph.D.s (all in mechanical engineering at Stanford University), these design-obsessed makers are devoted to the publication of *Ambidextrous* magazine, a quarterly that celebrates the people, process, and products of design. Check it out and/or subscribe at ambidextrousmag.org.

MacGyver Handbook

***The Unofficial MacGyver How-To Handbook: Revised 2nd Edition*, by Bret Terrill and Greg Dierkers** $20, American International Press

MacGyver always seemed like one of those "kids, don't try this at home" shows. Well, that was up until this spiffy book came out that includes a complete listing of every trick ever done on the show with step-by-step instructions for many of them. Checking in at about $20, you can get it for a reasonable price, and it actually makes for pretty good weekend project fodder (although disarming a missile with a paper clip is an academic exercise for most of us). Each Macgyverism includes a list of materials and a psuedo-scientific explanation of why it works; so much of it could be adapted without too much effort as a good middle-school science-class supplement.

—Matthew Russell

Design Yourself

***D.I.Y.: Design It Yourself*, edited by Ellen Lupton**
$20, Princeton Architectural Press

Finally — good design for the masses! This fabulously designed (of course) book was produced collaboratively by the self-described bastard children of the fine arts: MICA graphic design students and faculty. Based on the premise that "design is art people use" and should be accessible to everyone, the authors begin with basic design principles, in a writing style that feels like a good conversation.

Featured project ideas run the whole gamut from blogs, books, brands (fight the corporate power by making your own), and business cards (even chocolate brownie cards) to embroidery, flyers, housewares, logos, and yes, T-shirts. There's a great section on kids and design that highlights how design can serve as the best defense for kids against our corporate-sponsored landscape. Each page is an open invitation to reinvent your world, as well as yourself.

—Goli Mohammadi

Hitchhiker's Guide to the Workshop

Machinery's Handbook, 27th Edition $53, Industrial Press

Machinery's Handbook is the ultimate reference tool for makers and hardware hackers. In just the past few weeks, I've used it to look up a nice patina mix for copper artwork, calculate what thickness of steel I need to build a long-span desk, and figure out how many tons of sand my godchildren require in their sandbox. From A (ABEC bearings) to Z (zinc-based, die-casting alloys), the 70-page index alone will keep a hands-on person dreaming up new projects for a good long while. It's so much a part of hands-on culture that most machinist's toolboxes are manufactured with a special drawer that fits just one thing: this book.

—Travis J.I. Corcoran

Big Little Sharpie

Sharpie Mini $1 at office and drug stores, sharpie.com

I, like most women I know, have used the venerable black Sharpie for many wardrobe fixes — scuff marks and bleach splatters are no match for the queen of permanent markers. Then the colored Sharpies arrived, and oh, the possibilities. An eight-pack of various colors means I can hide the chips on the rim of my expensive platter and mask the gouges in my not-so-expensive kitchen cabinets. Pastels come out at Easter; red and green come out in December.

But none of them hold a candle to the new Sharpie Minis. Complete with a ring on the end so you can attach them to your keychain or backpack, they are maker nirvana in miniature. Now I always have a powerful writing utensil with me — I can copy something off a bulletin board; make adjustments to myriad items, from handbags to posters; or get an autograph on my son's T-shirt from his favorite musician. Permanent, colorful, stylish, non-toxic, utilitarian. Dreamy!

—Shawn Connally

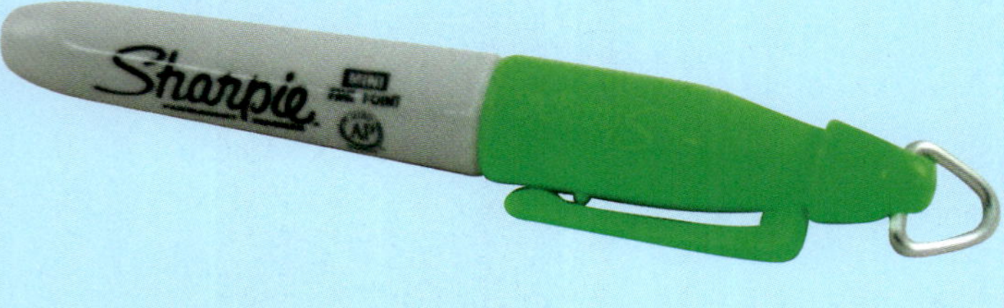

30 Amazing Minutes

***The Way Things Go*, by Peter Fischli and David Weiss** $20, amazon.com

You may have seen the Honda commercial where parts from an Accord are used to create a fantastic Rube Goldberg-like contraption. That footage pays homage to artists Fischli and Weiss' 1987 film, *The Way Things Go*, which opens with a large garbage bag suspended from a twisted rope. As the rope unwinds, the bag lowers until it bumps into a tire, which begins to roll, and the chain reaction begins its way through the warehouse.

Common objects are employed: tires, ladders, chairs, and bottles — even a humorous old pair of shoes. And there's more than just kinetic interaction: chemical reactions, fire, and even crude rockets and explosions are used to make things go. Sometimes events happen fast; other times, you have to endure the anticipation, perhaps while a container slowly fills with enough water to make a can float and tip a chair.

—Terrie Miller

See the Honda commercial at makezine.com/06/toolbox

Always Use Protection

Vinyl Film

Prices vary, available at craft stores

Five years ago, I wanted a clear protector for my iPod that worked like the sticker that Quickie-Lube puts on my car windshield. So I went to my local craft store to see what they had. I was initially intimidated by the different thicknesses of vinyl, but they let you decide if you want light protection from minor scratches or something that might absorb the impact of a drop onto a hard floor.

The thin stuff doesn't even require a cutout to let the click wheel move — it flexes just the right amount. I cut the material to size and wrapped it around my iPod like a burrito. It sticks really well to a clean iPod (and itself) without any adhesive. When my iPod died after five years of use, it didn't have a scratch on it.

—Dan Strunk

Shawn Connally is managing editor of MAKE.

Travis J.I. Corcoran hacks software, wood, and metal. He is founder and president of TechnicalVideoRental.com.

Steve Johnson (mojocode.com) is an IT consultant with a point.

Terrie Miller is online manager of MAKE.

Goli Mohammadi is copy chief of MAKE.

Tom Owad is the author of *Apple I Replica Creation* and editor of applefritter.com.

Dave Prochnow is a frequent contributor to MAKE. You can learn more about his projects at pco2go.com.

Joel Raedeke misses Ax-Man Surplus.

Matthew Russell tries hard to live life as a renaissance man, but is distracted by the cult of Mac.

Dan Strunk and his scratch-free iPod live in Minneapolis, Minn.

Phillip Torrone is senior editor of MAKE.

Have you used something worth keeping in your toolbox? Let us know at toolbox@makezine.com.

AHA! Puzzle This

By Michael H. Pryor

MAKE's favorite puzzles. (When you're ready to check your answers, visit makezine.com/06/aha.)

Coins on the Table

Your roommate challenges you to a game with the quarters from each of your sock drawers. Whoever wins the game gets to keep all the money. (Assume you each have an unlimited supply of quarters.)

You sit down at the perfectly round kitchen table and each person takes a turn, placing a quarter down anywhere on the table. No quarters can overlap and the entire quarter must rest on the table surface. The first person that can't put a quarter down on the table loses. You each have plenty of quarters and won't run out during the game.

Your roommate wants to go first. But you think about it for a while and realize you would much rather go first because you've figured out a surefire method to win (without cheating). After some convincing, your roommate allows you to place the first quarter.

Where do you place it, and what is your winning strategy?

In-Your-Head Math

How many trailing zeroes are there in 100 factorial (100!)? (100 factorial is 100 * 99 * 98 * etc., down to 1.) For example, 5,030,499,400,000 has five trailing zeroes.

One Mile South

How many points are there on Earth where a person can walk one mile south, one mile east, and then one mile north and end up in the same spot they started in? To be precise, let's assume the Earth is a solid smooth sphere, without oceans or mountains.

The North Pole is one such place where you can walk one mile south, one mile east, one mile north and end up where you started. But is there another starting point that works?

If you think you've figured it out, I'll give you a hint. There is more than one point. In fact, there are more than two points.

Calendar Cubes

Each morning I come into work and arrange two normal cubes on my desk to display the current day of the month. For example, on the 1st of the month, the cubes have a 0 and a 1 on them. On the 16th day of the month, one cube has a 1 and the other a 6. On the 31st day of the month, a 3 and a 1 appear on the front of the cubes. Each cube has six sides and one number painted on each side. What are the numbers painted on each cube?

Illustrations by Roy Doty

Michael Pryor is the co-founder and president of Fog Creek Software. He runs a technical interview site at techinterview.org.

READER INPUT

Where makers tell their tales and offer praise, brickbats, and swell ideas.

I've had to rewrite this letter in my mind several times to keep it from sounding like a fan letter penned by a silly schoolgirl to Elvis.

Wow! I just got my first issue of MAKE and I'm giddy with excitement. I'm a not-very-closeted geek, and have attempted (sometimes successfully) some fairly strange projects, but your people have shattered any illusions of grandeur I might have held. I read it cover-to-cover-to-cover and now I'm re-reading it again!

I'm reminded of the stories they tell about Nikola Tesla, how he had ideas, visions, of fully fleshed-out inventions flying through his mind so fast that he couldn't even snag them all as they raced by. Now after reading your magazine I think I may know just a little bit how he felt!

I've turned some like-minded friends on to this incredible ride, who now feel like they *need* to subscribe, simply by describing some of what I read in Volume 05. *—Dave Sica*

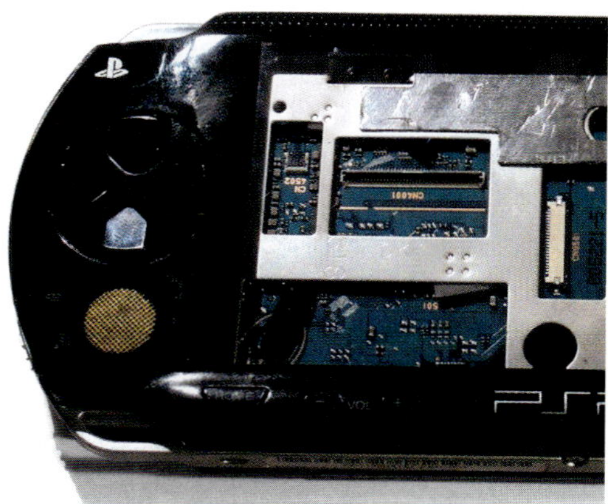

Hey, I love the magazine, keep it up. I was not all that happy with my PSP; I also attained an extreme distaste for Sony due to events following my purchase of the PSP. And so I was compelled to disassemble it. As soon as I broke the seal I knew that I had voided the warranty and so, I thought, why not proceed? In the process of putting it back together I decided to put the case back on and take this picture, and wanted to share it with you. I hope this makes you feel as good as it did me.

—SpotDart

MAKE AMENDS

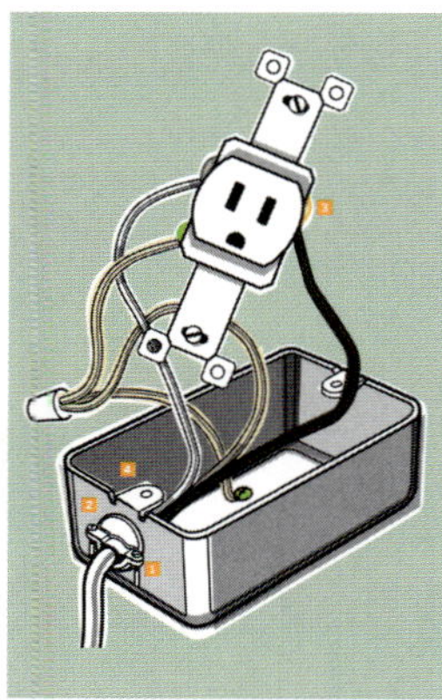

We heard quite a lot from our readers about the misdrawn illustration on page 112 of Volume 05. In the piece, "How to Build a Power Tap," the illustration can cause some confusion as well as a tripped circuit breaker. The drawing shows a dual plug receptacle, which, when wired improperly, can cause a short circuit. Do NOT attach the wires from the extension cord to the screws on the same side of the receptacle, because it will create a short circuit.

Many eagle-eyed online readers of our digital edition tipped us to our error, even before the print edition was mailed to subscribers or showed up on newsstands. This quick communication let us notify tens of thousands of subscribers via email before they even received their copies. It also allowed us to put a warning sticker on another 15,000 copies that hadn't yet shipped to retailers. Thank you for the help, and please know that we've added an electrical proofing step in our editorial process.

The corrected illustration is at left. You can also download it from makezine.com/images/05/p112revision.pdf. The PDF is designed so you can paste it right over page 112 in MAKE Volume 05.

OTHER ERRATA:

In Volume 05, page 58, "No More Scratches!" suggested using toothpaste to remove scratches from glass. To clear up confusion and avoid disasters, please remember that this ONLY works on glass (and should be tested in an out-of-the-way area first). It should never be used on plastics or on glass with coatings, and it will damage many modern eyeglasses, which, despite the name, are not made from glass.

In Volume 05, page 138, Fig. 2, the distance between the left edge of the base and the center of the leftmost mirror block was incorrectly labeled as 5⁄16" rather than 15⁄16". You can find the correct dimensions on author Rob Hartmann's blog, at pseudoscope.blogspot.com.

In Volume 05, page 175, the price of the Poqet PC was wrong. The Poqet PC actually costs $159 plus shipping.

BLAST FROM THE PAST

The Boy Mechanical

By Mister Jalopy

A look back at the glory years of engineering for common folk.

Being an insufferable snob is an awful lot of work. It can be heartbreaking, wallet-emptying, and full of regret — often all at the same time. In the world of snobs, the most pure would have to be the book snob. Sure, you can spend more money on Tiffany lamps and art from the New York School, but it is difficult to match the mania of a true-blue book snob. To winnow that group of individuals to an even narrower class of snobbery, consider the perfectionist bibliophile who buys books on eBay. While intoxicated. A surer recipe for regret has never been devised. My own disappointments are manifold — limitless potential and hope that nearly always end in disaster.

As a result of mixing eBay and Scotch, I own the original four-volume set of *The Boy Mechanic*, which was printed from 1913-1925. Naturally, I paid too much and was terribly disappointed with the condition of the books, but not with the content. Slack-jawed wonder lies within those tattered covers.

The projects range from the utterly pedestrian, like "How to Weave a Shoestring Watch Fob," to the ridiculously ambitious, as illustrated by the young lad who is gracefully flying his homemade glider off a cliff and over a locomotive only to land in his hometown. I bet the girls at his school were mighty impressed.

These were the glory years of engineering and mechanics for common folk. The Wright brothers had flown their airplane a few years earlier, the Model T was changing the world, and unlimited opportunity was waiting in the workshop. The insanely complicated projects, like the "Homemade Steam Turbine," are immensely entertaining, but are they any more enjoyable than the 44 parlor tricks, like "Removing 36 Cannon Balls From a Hand Bag"? Why, the toboggan made of homemade skis and an old ladder, in "Making Skis and Toboggans," looks to be every bit as good at knocking out front teeth as today's modern version!

Luckily, Volume 1 is available for free from Project Gutenberg, so you can dodge the high-price-and-poor-condition bullet that nearly killed me. Furthermore, if you are OK with having a PDF instead of a crumbly, yellowed, chipped volume from 1913, then you will have avoided the dreaded fate of the book snob.

The airborne boy.

Published in the early 20th century, *The Boy Mechanic* provides instructions on making "wireless outfits, boats, camp equipment, aerial gliders, kites, self-propelled vehicles, engines, motors, electrical apparatus, cameras and hundreds of other things which delight every boy."

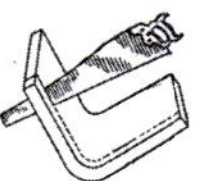

Making a boomerang from a "plank of well-seasoned hickory." It's noted that "any worker in wood can turn out a great number of boomerangs cheaply."

Mister Jalopy breaks the unbroken, repairs the irreparable, and explores the mechanical world at Hooptyrides.com.

Images excerpted from *The Boy Mechanic*, 1913

RETROCOMPUTING

Old Hardware: Anything but Garbage

By Tom Owad

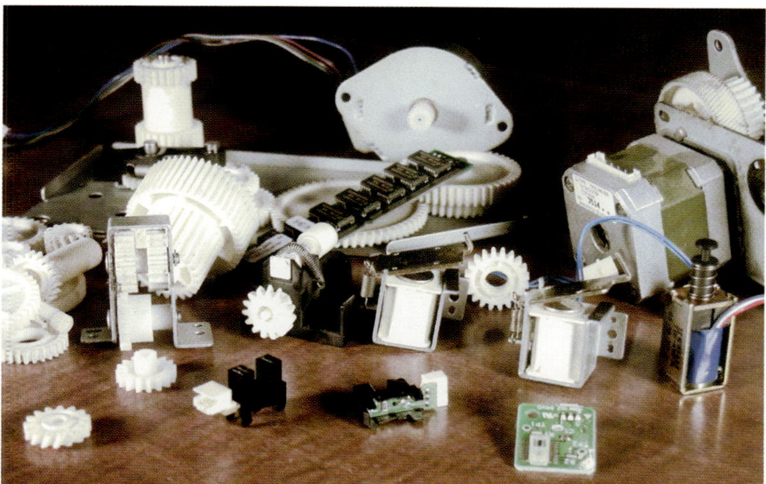

Laser printers and photocopiers are full of especially interesting mechanical parts, with myriad gears and a variety of sensors and electromechanical components.

When my Laserwriter 12/640 PS tragically passed away, I ended up with drawers full of useful and fascinating components.

Photograph by Tom Owad

Old computers vary greatly in their desirability. A handful of systems are in great demand. No Altair 8800 or DEC PDP-8 is going into the trash, but more common computers, especially ones that were popular in schools, are still headed for the dump by the truckload. What's to happen to all those Apple IIe classics, PC clones, and printers that can't even be given away?

I now tear apart anything I can get my hands on before passing it along to the garbage man. Rugged old computers are sure to have some quality parts within them, perfect for use in future projects. Any circuit board from the 80s is likely to have TTL or CMOS chips, and if you're lucky, they'll be socketed and take a second each to remove.

Many boards will also have EPROMs, which you can quickly identify by the sticker covering the chip's window. Remove the sticker, expose the circuit to ultraviolet light, and they're as good as new. Pick up an old EPROM burner at a swap meet or on eBay, and you can reprogram it.

Mechanical devices are even more likely to contain interesting parts. A typical scanner includes a stepper motor and a handful of belts and gears, plus a long stabilizing bar that guides the scanner head (and could be used to guide anything). The one I tore apart last week also had a nice transformer and an EPROM on its circuit board. Invariably, they'll have a fluorescent or xenon lamp, too, which can be made into a bench lamp, if nothing else. I haven't actually done this, but it would be fun to be able to point around to various objects in your house and say, "That lamp used to be my scanner; that night stand was my server; that digital picture frame is my old PowerBook ..."

Hard drives are such a frequent point of failure that I've yet to scrap a working one, but I've encountered plenty of dying drives — most of them have two extremely strong magnets in them. When I bought a magnetic CB antenna for my car, I was disappointed to find that it wouldn't stick to my trunk lid. I placed four magnets scavenged from dead hard drives on the inside of the lid, and they've successfully held the antenna in place at more than 80 mph.

5.25" floppy drives almost always contain a stepper motor to adjust the head and a DC motor to spin the disk. Scrap out a pallet of old PCs, and you'll have enough motors to build some pretty complex devices. The drives also typically have two stabilizer bars, just like in the scanners but smaller. I haven't yet figured out what I can make with these, but I've got a drawer full of them. 3.5" floppy drives tend to have junkier parts (especially the newer ones), but the motors are much smaller, which can be useful.

As systems get more compact, it gets harder to find decent parts, but it's still not impossible. The death of my wireless router yielded a nice collection of LED arrays, two crystal oscillators, an 802.11b card, and two antennas, complete with internal cabling.

Tom Owad (owad@applefritter.com) is a Macintosh consultant in York, Pa., and the editor of applefritter.com. He is the author of *Apple I Replica Creation* (Syngress, 2005).

Thank You to Our Sponsors

TALES FROM MAKE: BLOG

Roomba Tronic

By Phillip Torrone

Take a tour into the underground robotic relentlessness of Roomba hacks, robots in the streets of Austin, and robot cockfighting.

Roomba wars!

It all started out innocently enough. iRobot recently opened up their Roomba vacuum for hackers, tinkerers, and makers to send control signals via a little hidden port (irobot.com/hacker). On Makezine.com (makezine.com/blog), we posted how to control your Roomba with a serial cable or a Bluetooth computer, and all was well in the world. But you simply cannot have robots under human control; they're better off sticking to Isaac Asimov's "Three Laws" than listening to us meat-based guardians. With great robot power comes great robot responsibility.

At O'Reilly's yearly Emerging Technology conference, alpha geeks and technologists from around the world gather to tune in, turn on, and trade RSS feeds. One of the nightly events, the MAKE Fest, showcased a dozen makers dutifully explaining their creations, from wired clothing to retro arcade mods.

This is when it happened.

The ambient noise of conference chatter stopped, cheers were heard, money was thrown on the table. An underground Roomba cockfighting competition spontaneously erupted. Maker Tod Kurt (todbot.com) just happened to have his Roomba with him; we happened to have a couple extra with us (you just gotta love a conference where people have extra Roombas with them).

Bets were taken; weapons were formed from what we had available (mostly conference schwag). The crowd frenzied; mild-mannered AJAX programmers were pushing their way in to get a piece of the robot action. Oh yes, it was on.

A few days later the MAKE team was off to Austin's South by Southwest Interactive Festival. Our little MAKE blog picked up two awards for Best Educational Resource and Best Craft Blog, and in celebration we continued our relentless pursuit of robot gaming. The details are hazy, but basically

With great robot power comes great robot responsibility.

we hustled a bar full of festival goers with some Roomba pool. And still later, with the help of some Austin residents, we played the first game of "Real Frogger." This in celebration of Frogger's 25th anniversary, of course.

The future of robotic games has arrived, it just hasn't been evenly vacuumed. Complementing this issue of MAKE (our robotics theme), check the makezine.com/blog each day; we'll have new projects, updates, and more things you should (and shouldn't) do with robots.

Phillip Torrone is associate editor of MAKE.

Photograph by Scott Beale, Laughing Squid / laughingsquid.com

MAKER'S CALENDAR Compiled by William Gurstelle

Our favorite events from around the world — June - July - August 2006

June

»The Power of DC June 3, Hagerstown, Md. Fast, powerful, and silent electric vehicles race for ecologically sound glory. nedra.com

»Combine Demolition Derby June 10, Lind, Wash. Lind's yearly rodeo weekend kicks off with a "Combine Demolition Derby" showcasing kamikaze farm machinery. lindwa.com

EDITOR'S CHOICE

»Made in America Factory Tour Weekend June 14-17, York, Pa. Visitors are invited onto the production floors at companies such as Harley-Davidson, Pfaltgraff, Utz, and nearly 30 others. factorytours.org

»RoboGames San Francisco State University, June 16-18, San Francisco. Robotics galore: FIRA, BEAM, Mindstorms, and Robotic Combat. robogames.net

EDITOR'S CHOICE

»LDRS 25 Silver Anniversary June 28-July 4, Amarillo, Texas. The largest amateur rocketry event in the world, featuring hundreds of high-powered rocket launches. ldrs.org

July

»Montréal International Fireworks Competition Weekends, June through July, Montreal. Teams from around the world compete in what may be the largest and most prestigious pyrotechnical competition anywhere. internationaldesfeuxloto-quebec.com/en

»Warship Weekend 2006 July 8-9, The Museum of Naval Firepower, Gosport, UK. Display features old and modern warships. explosion.org.uk

»Da Vinci Days July 14-16, Corvallis, Ore. A multidimensional festival of Da Vinci-esque activities, including a kinetic sculpture race, robots, speakers, and more. davinci-days.org/index.php

»The Royal International Air Tattoo July 15-16, Fairford Air Field, Fairford, UK. RIAT is the biggest military air show held in Europe and attracts participants, airplanes, and technology from around the world. airtattoo.com

»ASME's Human Powered Submarine Contest July 19-23, Escondido, Calif. Engineering students will test their skills by designing, constructing, and racing their own submarines. sections.asme.org/sandiego/HPS2006.htm

»World Freefall Convention July 21-30, Quincy, Ill. The USA's largest skydiving convention occurs every summer at this central Illinois location. freefall.com

»Farnborough International Airshow July 22-24, Farnborough, UK. This huge airshow features a 4+ hour flying display, static displays, exhibitions, seminars, rides, and other attractions. farnborough.com

»The 2006 Fire Arts Festival July 12-17, The Crucible, Oakland, Calif. Pyro-artistic festival includes lectures, classes, sculptures, demonstrations, and fire-related performances. thecrucible.org/calendar

»EAA AirVenture July 24-30, Oshkosh, Wis. More than 750,000 people attend AirVenture, the annual convention of the Experimental Aircraft Association with thousands of aircraft-related events. airventure.org

August

»DEFCON August 4-6, Riviera Hotel, Las Vegas. DEFCON is one of the largest and oldest computer hacker conventions in the world. defcon.org

»6th Annual Telluride Tech Festival August 12-14, Telluride, Colo. The festival looks at the past and future of technology, through speakers and discussion groups. telluridetechfestival.com

EDITOR'S CHOICE

»Bonneville Speed Week August 12-18, Bonneville Salt Flats, Utah. The fastest racers on Earth streak across the saltpan attempting new speed records. scta-bni.org

Important: All times, dates, locations, and events are unconfirmed and subject to change. Verify all information before making plans to attend.

Do you know of an event that should be included? Please use our online form at ***makezine.com/events****. Sorry, it is not possible to include listings for all events in the magazine, but they will be listed online.*

*Did you attend an interesting event? If so, please tell us about it in our Forums. (**forums.makezine.com**).*

GEORGE DYSON

Off-the-Shelf Parts, Off-the-Wall Ideas

Looking for the dawn of the digital universe? Check in the basement, next to the lavatory.

Sixty years ago, in the spring of 1946, a handful of electronic engineers showed up at the Institute for Advanced Study in Princeton, N.J., and requested a place in the basement to plug in their tools. They were out of luck. "The only really useable space in our basement is that adjoining the men's lavatory, to which you are most heartily welcome," Electronic Computer Project director John von Neumann was told. Despite this inauspicious welcome, the engineers moved in, and in April, the project's accountants recorded the first $4 for "electrical work."

The Institute for Advanced Study was established in 1930 as a refuge for scholars in mathematics, physics, history, and art. There were no laboratory facilities of any kind. "What could be wiser than to give people who can think the leisure in which to do it?" the founding director Abraham Flexner had been advised in 1939. Not a word about workbenches or tools.

John von Neumann was a pure mathematician who, along with many of his colleagues, had been drawn over to the applications side of things during World War II. He liked it there, and refused to leave. Having been exposed to the powers of digital computing through his involvement with the nuclear weapons program at Los Alamos, he was impatient to build the next generation of computers, and was well aware of the implications of Alan Turing's 1936 results on Universal Machines. "We are building one," he told his arriving engineers, and the copy of Turing's paper in the IAS library has been so heavily consulted that its binding has come unglued.

Other groups were engaged in similar projects, but none quickly and boldly enough for von Neumann. "If he really wanted a computer," explained Arthur Burks, "the thing to do was to build it."

Burks and Herman H. Goldstine, an Army colonel transplanted from the ENIAC project at the Moore School in Philadelphia, were installed on the second floor of Fuld Hall in a small office belonging to logician Kurt Gödel, where they set to work, under the guidance of von Neumann, outlining the logical architecture of the machine.

"The heart of the system is a central clock, carrying an enormous load," say the minutes of the first meeting of the Electronic Computer Project. The circuitry would be modular, because "this sort of design is favorable for mass production," added the engineers. "Words coding the orders are handled in the memory just like numbers," explained von Neumann, breaking the distinction between numbers that *mean* things and numbers that *do* things, ensuring that all hell would break loose.

Instead of importing a few mathematical logicians into a lab full of engineers, von Neumann imported a handful of engineers into a place run by mathematicians. Although they had to start out by building their own workbenches at the Institute, and scrounge for war-surplus components as best they could, there was no one around to say, "Well, Dr. von Neumann, this looks interesting, but we build computers like this ..."

The lead engineer was Julian Bigelow, former colleague of Norbert Wiener and co-author of the 1943 paper "Behavior, Purpose, and Teleology," around

> Every single bit of the digital universe we now inhabit can be traced back directly to those first binary digits that flickered to life in the summer of 1951.

which the beginnings of the cybernetics movement coalesced. His job was to take the logical design as laid out in the abstract by Burks, Goldstine, and von Neumann, and coax it to life as a machine. "

Julian would have the ideas, and Ralph Slutz would kind of detail the ideas, and then James Pomerene and I would go try and make the electrons do their thing," explains Willis Ware, one of the original engineers. "Julian was the architect of that machine." Bigelow's gang soon outgrew their room in the basement of Fuld Hall, and were pushed outside to a hastily constructed building at the edge of the Institute's 600-acre woods.

At the Institute for Advanced Study, it was easier to find an expert in quantum mechanics than someone who worked on their own car. Bigelow was the exception — the only permanent member of the

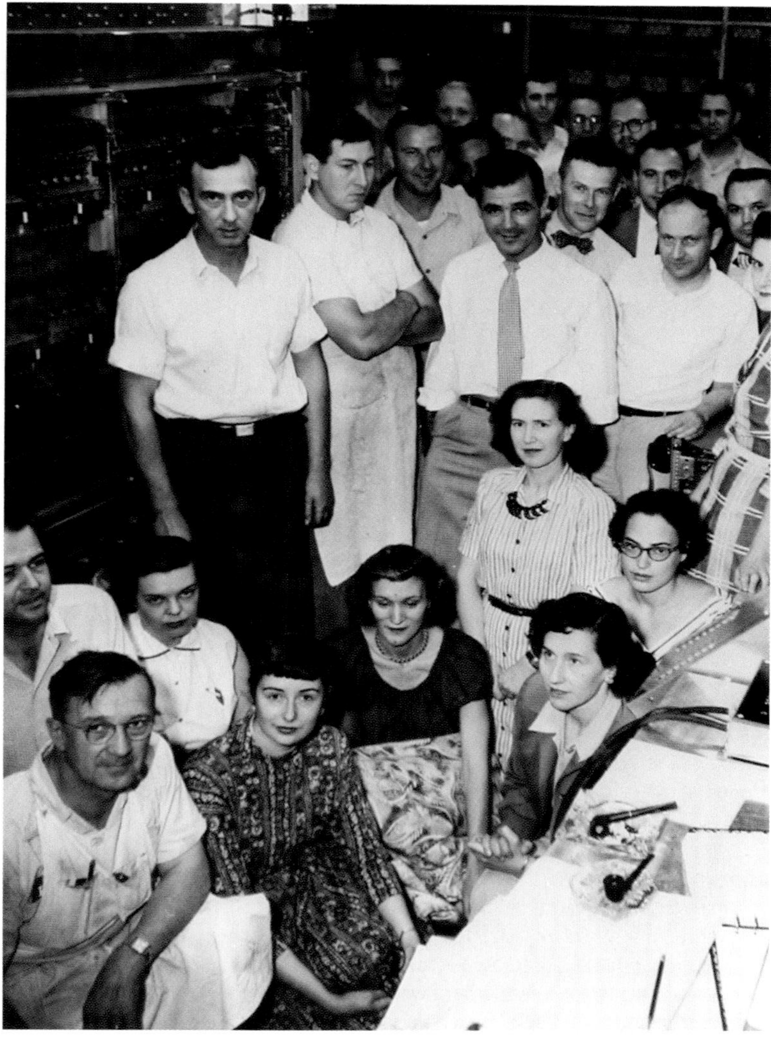

Making a Revolution: Staff of the Institute for Advanced Study Electronic Computer Project, ca. 1952. After John von Neumann decided to construct a modern, stored-program, high-speed electronic digital computer, others had to design and build the machine, and write the code to make it work. Some of these people, incompletely identified, are pictured here.

The entire computer was built in the project's primitive machine shop, and all code was assembled by hand, using absolute addressing, bit by bit. The finished computer, comprising a total of 3,474 vacuum tubes, had an internal memory of 40,960 bits, stored as 32x32 arrays of charged spots on the faces of 40 cathode-ray "Williams Tubes." It entered service during the summer of 1951 and was duplicated widely around the world.

Photograph by Alan Richards, courtesy of the Archives of the Institute for Advanced Study

Institute who routinely dismantled engines with his own hands. He approached the construction of the new computer as a true engineer. The guts of the new machine (the first to have a substantial, fully random-access memory available at megacycle speeds) were 40 "Williams Tubes," 20 on each side of the main chassis, so that the computer not only resembled, but functioned, as an enormous V-40 engine with overhead cams and valves.

Every single bit of the digital universe we now inhabit can be traced back directly to those first 40,960 binary digits that flickered to life, in the summer of 1951, as a 32x32x40 matrix of charged spots on the face of some delicately modified cathode ray tubes.

The first serious computation, a thermonuclear problem for Los Alamos that was coded by a crew including Hedy Selberg and Klari von Neumann, ran for six weeks nonstop, and led to the success of the hydrogen bomb. Not bad for a shakedown run.

Cosmologists may never know what existed before the big bang, but at the dawn of the digital universe, we know that it was Julian Bigelow, James Pomerene, Herman Goldstine, Ralph Slutz, Jack Rosenberg, Willis Ware, and a small group of helpers who worked to bring the dreams of Leibniz, Babbage, Turing, Atanassoff, Zuse, von Neumann, and many others to life with their bare hands.

George Dyson, a kayak designer and historian of technology, lives in Bellingham, Wash., and is the author of *Baidarka, Project Orion,* and *Darwin Among the Machines.*

My Version of Einstein's Amplifier

By Tyler Rourke

After World War II, many of the best physicists and mathematicians in the world could be found at the Institute for Advanced Study (IAS) in Princeton, N.J. In 1949, a young engineer named Jack Rosenberg was part of a team building one of the world's first digital computers. Having knowledge of electronics, a keen ear, and an interest in classical music, Rosenberg's home-built hi-fi was far superior to commercially available audio systems, and before long, his system was widely talked about at the IAS.

Erwin Panofsky, an art historian and close friend of Albert Einstein's, heard about Rosenberg's amazing hi-fi and asked him if he would build a system for "The Professor" as a 70th birthday gift from everyone at the IAS. Rosenberg agreed, and on March 14 he and J. Robert Oppenheimer personally delivered the system to Einstein at his home.

Last summer, I found myself speaking with George Dyson, who had spent time during his youth at the IAS. He told me that he had recently interviewed Rosenberg about the hi-fi. He offered a copy of the original hand-drawn schematic of the amplifier portion of Einstein's hi-fi system and encouraged me to attempt a recreation of it. I immediately began figuring out how I was going to build the amplifier.

I started the project by purchasing a chassis that was similar in size to the one specified in the parts list. Although I was unable to obtain the original transformers, I found suitable substitutes and positioned them on the chassis. I then laid out the positions of the tubes and tried to arrange all of the parts (switches, jacks, pilot light, etc.) according to function, so that high voltages would be isolated from the signal path whenever possible. I printed templates to mark the positions of all holes, drilled the chassis, installed the hardware, and began laying out the amplifier's internal components. Resistors and capacitors were connected directly to tube sockets, switches, and jacks wherever possible. The remaining components were soldered to terminal strips, which were then connected according to the schematic. Friends at CDI Signs in Bellingham, Wash. made the beautiful faceplate for me.

Visually, my version is not identical to the original, and I've had to substitute some parts that were no longer available, but I'm confident that this amplifier sounds nearly the same as Einstein's, which is a wonderful thing.

Tyler Rourke has a penchant for electronics and good food.

Photograph by Tyler Rourke (thanks to Jack Rosenberg)